AF757335

SYSTEMIC REGULARITIES AND SYSTEMIC OPTIMIZATION

Systemic Regularities and Systemic Optimization

Iveri V. Pranguishvili
Vladimir N. Burkov
Ivan A. Gorgidze
Guram S. Javakhadze
and
Ramaz A. Khurodze

Nova Science Publishers, Inc.
New York

For permission to use material from this book please contact us:
Telephone 631-231-7269; Fax 631-231-8175
Web Site: http://www.novapublishers.com

Library of Congress Cataloging-in-Publication Data

Systemic regularities and systemic optimization / Iveri V. Pranguishvili ... [et al.].
p. cm.
Includes bibliographical references and index.
ISBN 978-1-60741-869-6 (hardcover)
1. Organizational behavior. 2. Decision making. 3. Statics and dynamics (Social sciences) I. Pranguishvili, Iveri V.
HD58.7.S966 2009
302.3'5--dc22

2009017711

Published by Nova Science Publishers, Inc. ✣ New York

The book is dedicated to the 30-anniversary
of Enthronement and the 75- birthday
of His Holiness and Beatitude,
Catholicos-Patriarch of All Georgia Ilia II

CONTENTS

PREFACE

Any organizational system seeks to improve (optimize) the efficiency of its activities in terms of achieving the goals it faces. Sometimes this optimization is conducted by the intrinsic forces of the system itself (self organization; however, sometimes it is conducted by external forces interested in improving the efficiency of its functioning, and sometimes it is conducted by both intrinsic and external forces. The scientific book "Systemic Regularities and Sytemic Optimization" (I. Prangishvili, V. Burkov, I.Gorgidze, G. Javakhadze, R. Khurodze) describes effective approaches to solve the tasks of systemic optimixation. The authors state that the process of improving efficiency of an organization can be in less detail shown as consisting of the three stages. For each stage, special methods and informational technologies are applied. The first stage comprises the elaboration of the development (improvement) strategy for the organization. At the second stage according to the strategy elaborated at the previous stage the authors observe the change (optimization) of the structure and parameters of the organizational system, which determine its functioning (i.e. specify the range of possible trajectories of its development). At last, at the third stage a specific trajectory (plan) of its development is determined. All three stages together are called systemic optimixation, "Systemic Regularities and Systemic Optimization" is recommended to management specialists, and, also, can be used for teaching, professional reference, etc.

Chapter 1 - In this chapter the authors show that major objective natural and social systemic regularities determine the scenario of the future social and economic development of the world community and that no universal progress and prosperity like communism can ever take place in the life of all the peoples in the world because it would upset the major systemic regularity of the entropic balance. Below the authors analyze systemic regularities which enable us to understand the reasons for provoking social conflicts, initiating military action and natural catastrophes. On the basis of objective regularities an attempt can be made to predict the development of Russia in the twenty-first century. The objective systemic regularity of the four-stage spiral development makes it possible to determine the main reasons for changing polities (dictatorship, democracy, anarchy). Besides, it becomes possible to determine reasons for treachery of business partners and infidelity of sex partners [22].

Chapter 2 - This chapter is dedicated to describing the tasks for optimizing distributed decision-making systems [10, 15, 33]. Regional development programs [6, 15, 24] are considered as an applied model. Such programs represent a complex of interrelated activities aiming at achieving certain development goals which are estimated according to social, economic and other criteria. The programs include a set of projects, and projects include sets of subprojects. Typical examples of such projects are projects for reforming and restructuring (PRR) enterprises [23], organizations, their subdivisions and other objects directed at improving the operating benefits of the object undergoing a reform. As a rule, each PRR helps to achieve simultaneously a number of goals and requires certain costs. Therefore, operating benefits may be estimated according to a number of parameters. Accordingly, the effect achieved as a result of PRR implementation is estimated according to several criteria. The management problem consists in choosing a great number of PRR-s to be implemented and supported in the first place under the input restraints.

Chapter 3 - The theoretical results gained in the organizational systems control theory [8, 10, 20] are used in making applied models which in turn are practically used in synthesis or modification of control mechanisms for real social-economic systems. From the point of view of the scale, regions are the largest control objects. In elaborating and implementing regional development programs the following methods are used: methods for a complex estimation of regions, competitive mechanisms for selecting enterprises and including them in the regional development program, methods for optimizing programs according to their cost, mechanisms for distributing financial resources including consent mechanisms and expert mechanisms [1, 2, 4, 7, 8, 11, 15, 16]. This chapter deals with a number of models and methods for systemic optimization of regional development programs (it implies a number of interconnected actions aimed at achieving the region development goal which, as a rule, are estimated according to a number of criteria) [7].

Introduction

Any organizational system seeks to improve (optimize) the efficiency of its activities in terms of achieving the goals it faces. Sometimes this optimization is conducted by the intrinsic forces of the system itself (selforganization), however, sometimes it is conducted by external forces interested in improving the efficiency of its functioning, and sometimes it is conducted by both intrinsic and external forces.

The process of improving the efficiency of the organization can be in less detail shown as consisting of three stages. The first stage comprises the elaboration of the development (improvement) strategy for the organization. At the second stage according to the strategy elaborated at the previous stage we observe the change (optimization) of the structure and parameters of the organizational system, which determine its functioning (i.e. specify the range of possible trajectories of its development). At last, at the third stage a specific trajectory (plan) of its development is determined. All the three stages together are called systemic optimization.

From the point of view of temporal periods the first stage is the most inertial, it determines the guiding line and priority directions for the development of the organization (the development strategy). A development strategy for a country is usually elaborated for 15-20 years, and for a firm – for 3-5 years. The second stage (choice of the structure and structural parameters) for the country covers 3-5 years, and for a firm – 1-2 years. At last, the third stage is the least inertial (it usually covers 1-2 years for a country, and a month or a quarter - for a firm).

Special methods and informational technologies are applied at each stage. At the conceptual stage it is very important to take into account systemic laws and systemic regularities. Unfortunately, in organizational systems the laws are largely philosophic and quite general. Quite different are systemic regularities, i.e. certain relationships between the system parameters, which in a manner are probabilistic (with quite a great degree of probability it can be stated that they will take place). Systemic regularities represent a generalization of historic experience concerning the development of numerous systems. The knowledge of systemic regularities makes it possible to avoid blunders at the conceptual stage and to enhance the validity of the elaborated strategy.

At the second (structural) stage structural optimization methods are applied, and at the third one optimal control methods are used.

An attempt to describe a number of approaches to solving the tasks of systemic optimization is made in this book.

The first chapter deals with main systematic regularities and provides examples of applying them for decision-making at the conceptual level.

The second chapter is mainly connected with methods of structural optimization by giving an example of synthesizing distributed decision-making systems in regional systems. And at last, the third chapter deals with optimization methods for the functioning of organizational systems analyzed by giving examples of regional development tasks.

The first chapter is written by I.V.Prangishvili, the second chapter – by V.N.Burkov, I.A.Gorgidze and the third one – by G.S.Javakhadze, R.A.Khurodze and I.A.Gorgidze. D.A.Novikov's materials were amply used in writing the second and third chapters.

Chapter 1

GENERAL SYSTEMIC REGULARITIES

1.1. INTRODUCTION

We will make an attempt from the systemic standpoint to receive answers to some essential questions, generated by nature, technology, and society.

Thus, for instance, it is important to understand whether the world and people are governed by chance or by some objective systematic natural and social regularities. Below it is shown that certain systematic regularities represent the major determinants, and the degree of randomness being rather low, still introduces some amendments to this process.

In this chapter we show that major objective natural and social systemic regularities determine the scenario of the future social and economic development of the world community and that no universal progress and prosperity like communism can ever take place in the life of all the peoples in the world because it would upset the major systemic regularity of the entropic balance. Below we analyze systemic regularities which enable us to understand the reasons for provoking social conflicts, initiating military action and natural catastrophes. On the basis of objective regularities an attempt can be made to predict the development of Russia in the twenty-first century. The objective systemic regularity of the four-stage spiral development makes it possible to determine the main reasons for changing polities (dictatorship, democracy, anarchy). Besides, it becomes possible to determine reasons for treachery of business partners and infidelity of sex partners [22].

In this chapter we consider only some 28 objective general-system or universal regularities[1] studied in [22]. These regularities make it possible to give positive answers to the above-mentioned questions.

All systemic laws and regularities, as a rule, are of limiting nature and warn us what we should not strive for and thus have a guiding influence on our activities. If we know what the fundamental limiting nature of systemic laws and regularities consists in, we will better see the way towards what is really possible and what is impossible and as a result we would draw up guidelines, first of all, in the sphere of systems management. Thus, for example, it is known that both laws of thermodynamics eliminate a possibility of producing the first and second types of a perpetuum mobile or perpetual motion machine [8].

---

[1] In the given work by the term "regularity" we understand interconnection of phenomena (properties), which unlike in laws becomes apparent in the majority of cases, however there are some exceptions.

The knowledge of general systemic regularities and taking account of them makes it possible to reveal people's incorrect actions and errors in case their administrative actions contradict the existing natural and social regularities, especially while controlling complex semistructured and semiformalized technical, ecological, social-economic and political systems.

In addition, taking account of systemic regularities makes it possible to foresee processes beginning in crisis situations and define the most effective ways of withdrawal from them or ways to avoid them.

Besides, general systemic regularities help by analogy and isomorphism to ensure the shift of knowledge concerning major processes, taking place in complex systems, from one (well studied) sphere to another (less-studied) one regardless of their nature [22, 23].

Taking account of general systemic or universal regularities helps to find out possibilities and directions of development (improvement and organization) or degradation of different complex systems and, besides, to understand processes which begin in a crisis situation and determine effective methods of controlling various complex systems.

It will be shown below that on the basis of one (of 28) general-system regularities and, specifically, the regularity of entropic dynamic balance, or of increase or decrease of entropy in open systems the conditions of organization, structurization or selforganization and disorganization of various systems are determined. In the processes of structurization (selforganization) and disorganization taking place in the system a crucial role is played by the influence of environment. It will be shown that entropic balance between order (organization) and disorder (disorganization) ensures the stability of the system. It is also shown that for ensuring the stability and progressive development of the system it is necessary to learn to control: 1) the amplitude and frequency of entropic fluctuations; 2) the level of entropic balance and 3) take excessive entropy out of the system (transfer it to other systems or to the universe).

The oscillating and cyclic general-system regularity makes it possible to forecast the future development of Russia in the twenty-first century and in the third millennium.

The regularity of the dependence of the system's potential on its structure enables us to distinguish systems from pseudosystems.

The background regularity enables us to discover and identify unobtrusive objects by the change of the background radiation.

Before we proceed to stating universal or general-system regularities, we will briefly examine some fundamental principles concerning entropy, which are useful for further development of the problem.

1.2. Entropic Factor and Entropy Compensation Principle

Any systems of wildlife and abiocoen without exception comprise features of entropy and antientropy (or negentropy), i.e. features of disorder (chaos) and order, determinacy and indeterminacy, organization and disorganization.

The entropic factor as a quantitative measure of indeterminacy, disorder, disorganization (chaos) is universal.

We live in the world of molecular disorder and macroinstability and, therefore, we accept entropy as a universal parameter. Ensuring entropic balance between order and disorder in natural and social systems determines the behavior of our universe, world processes in space and conditions for life and development on our planet.

In scientific literature today at least the following four forms of entropy are found :

- the first form of entropy is considered was a degree of indeterminacy of behavior of any system including living and nonliving objects and their functions. Just this form of entropy as a degree of indeterminacy, which is represented by the number of acceptable state (S) of the system ($E \equiv \ell nS$), or as standard of a system behavior, degree of disorder, chaos, disorganization has become widespread of late and in this work we will use entropy exactly this sense;
- the second form of entropy is understood as a degree of indeterminacy of information which gives data concerning a certain system;
- the third form is represented by entropy of a molecular (microscopic) set;
- the fourth form is the entropy or indeterminacy of the state of any not quite ordered (arranged) system up to microscopic sets.

Alongside with usual positive entropy in nature there occurs antientropy or negative entropy, i.e. entropy with a negative sign. A system, as a rule, possesses antientropy (negative entropy), when there does not exist a complete procedure or algorithm for reproducing it.

A certain limited list of systems and processes of positive, negative and zero entropy bearers is given in Table 1.1 borrowed from [8].

Table 1.1.

Type of entropy (positive, zero and negative)	Type of order	Some systems and processes possessing positive, zero and negative entropy
Positive entropy or just entropy (E)	Indeterminacy, disorder (partial or complete chaos), disorganization	Molecular or macrosystems including living organisms, psychic abnormality, sometimes sound mind, information
Zero entropy (E = 0)	Complete order, complete determinacy, lack of chaos, organization	Logical thinking, logical conclusions, logical decision, sound mind, intellect, intuition, lack of information
Antientropy or negative entropy (- E)	Selforganization, complete selfcontrollability, selfdevelopment	Human brain and mind, living organisms as a whole

As a rule, all systems require repairing and rehabilitation after which their entropy decreases. Therefore, it is necessary for each system to have service personnel who periodically, after each process of rehabilitation decrease the entropy of the system. A human brain with its neural network is an exception. It does not need any outside service which would periodically decrease the brain entropy by means of repairing and prophylaxis. This fact makes us think that nerve cells themselves are bearers of necessary antientropy (negative entropy), and their death results in emitting antientropy necessary for intellect and that permanently dying cortex neurons serve as "fuel" for the power of thought. It is supposed that impulses of dying neurons are transferred to quanta of negative entropy (antientropy), therefore, human brain can on average produce up to one hundred thoughts a second [8].

It is impossible to explain the relations in the system between the substance (mass), energy and information because of the difference in their energetic potentials by a factor of 21 if not to admit in addition to substance and energy the existence of some additional mode (field) of objective reality – negentropy. Negentropic field or integrated information exists at all the levels of system hierarchy. If the substance (mass) of a system is connected with the field of mechanical and gravitational forces, and the energy of the system is connected with electromagnetic and temperature fields, then the negentropy of the system is connected with a negentropic field which is a bearer of information. As information, being a non-material process, cannot directly influence the substance and energy of the system and change it, the information received by the system, affects the negentropy of the system, and the negentropy in its turn is connected with the energy and mass. Negentropy defines the presence of both integrated information and structural order in the system. Therefore, systems are manifested by means of energy, substance (mass) and negentropy. Different systems influence each other by means of exchanging energy, substance and negentropy. Either energy, substance, or negentropy (information) may dominate in each separate system.

For general visualization of change – decrease of entropy of an open system we will consider interaction of an open system with the environment and the influence of the latter on the system. Open systems exchange substance, energy, information with the environment and thus change their structure and accordingly both sides increase their order, organization and decrease their entropy. This may be exemplified by all the effects of selforganization in wildlife and abiocoen, biological evolution, the origin and development of human society. Open systems receive energy, substance, information from the environment and thus decrease their entropy and increase the order and organization of their structure mainly at the expense of increasing entropy of the environment.

Indeed, under any change of the open system's condition the change of its entropy dE (or ΔE) is observed. It consists of two components: $dE = dE_1 + dE_2$, where dE_1 is the change of the entropy of the system resulting from the fact that the system exchanges (receives) energy, substance, information with the environment, and dE_2 is the change of entropy due to the processes taking place inside the system without any impact from the environment. If change dE_1 caused by the influence of the environment through energy, information, and substance is directed towards decrease of nonequilibrium, creating new resources, then change dE_2 is always spontaneously directed towards its growth, equilibrium. For irreversible processes always $dЭ_2 \geq 0$, and for reversible processes $dE = 0$.

Unlike open systems, closed or isolated systems do not exchange with the environment and, therefore, for them $dE = dE \geq 0$, which means that entropy of closed or isolated systems is irreversible and cannot decrease, but can only increase which gradually leads to the collapse of the original structure of these systems as a result of increase of indeterminacy and chaos caused by the growth of entropy.

Since under the second law of thermodynamics for any systems (including open ones) $dE_2 \geq 0$ occurs, general decrease of entropy and increase of organization of open nonequilibrium systems are possible only at the expense of the component $dE_1 < 0$. Thus, for example, a living organism can remain living only by permanently extracting negative entropy from the environment and, thereby increasing its positive entropy. It is obvious that to increase the degree of organization of the open system they must be nonequilibrium and, therefore, $E < E_{max}$. It is known that nonequilibrium systems are more organized, than equilibrium ones and their entropy is lesser.

As a universal parameter of the condition of a system sometimes instead of entropy (E) we use generalized entropy (GE), which is caused by the fact that we live in the world not only of molecular disorder, where entropy (E) is used as a parameter of condition, but also in the world of macrodisorder and macroindeterminacy of complex systems when it is reasonable to use generalized entropy instead of entropy. However, in what follows for the sake of simplicity we will use E instead of GE for describing disorder and indeterminacy even in complex systems. At the same time it should be noted that the absolute meaning of negative entropy of a system differs from its negentropy (NE).

Entropy (E) of the system as a quantitative measure (degree) of disorder of a system is defined by the number of admissible conditions of the system (S). Therefore, entropy is defined as a natural logarithm of a number of admissible conditions, i.e. $E \equiv \ln S$. The greater is the number of admissible conditions, the greater is the entropy of the system. Under $S = 1, E = \ln 1 = 0$, and under $S = S_{max}$, $\mathrm{E} = \ln S_{max} = E_{max}$.

An isolated or closed system A_3 consisting of two closed or isolated systems (nonengaging) systems A_1 and A_2 (Figure 1.1.) will have $S_1 \cdot S_2$ number of admissible conditions, where S_1 - the number of admissible conditions of system A_1, and S_2 - the number of admissible conditions of system A_2. Then $E_3 = \ln(S_1 S_2) = \ln(S_1) + \ln(S_2) = E_1 + E_2$.

Any closed or isolated system (A_1, A_1, A_3) strives for equilibrium condition when the number of admissible conditions of the system is maximal (S_{max}) and, consequently, their entropy is maximal as well (E_{max}). More exactly, entropy of any isolated system with overwhelming probability in due course will increase or at least will remain constant, i.e. $\Delta E \geq 0$ [7].

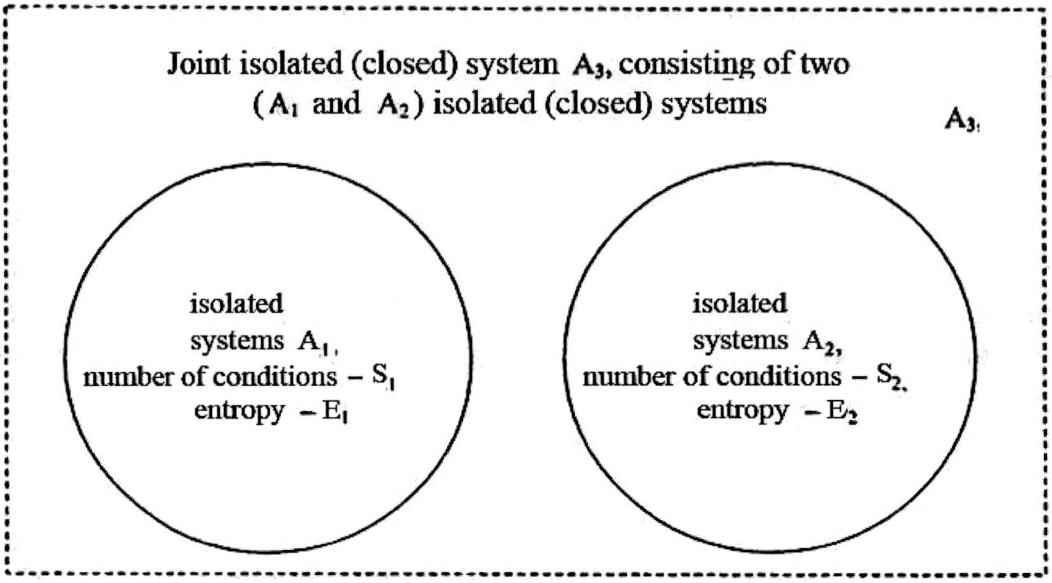

Figure 1.1. Two isolated (closed) systems A_1 and A_2 create a third isolated (closed) system A_3.

Now let us consider the closed or isolated system A_3 which does not interact with the environment or other systems and consists of two open or interacting (not isolated) systems A_1 and A_2 (Figure 1.2). For such an isolated (closed) system consisting of interacting systems there exists an entropy “compensation principle” which states that “in a non-isolated, and consequently in an interacting system the entropy of one system may be decreased if in the process of interaction with another system or with other systems there occurs a compensatory increase of entropy of another system” [25]. Then for two open and interacting systems A_1 and A_2 placed inside an isolated (closed) system A_3 the following expression for changing entropy $\Delta E_3 = \Delta E_1 + \Delta E_2 \geq 0$ is valid. It follows from this expression inside isolated system a A_3 is taking place in one system (for instance, A_1) decrease of entropy $\Delta E_1 \leq 0$, but another (A_2) the entropy must increase $\Delta E_2 \geq 0$ as much as it decreased in order to satisfy the condition $\Delta E_3 \leq 0$ for isolated system A_3. Indeed, entropy E_3 of a general isolated system consisting of two interacting (non-isolated) systems A_1 and A_2 (Figure 1.2) will with overwhelming probability increase or at least remain constant $\Delta E_3 \geq 0$, which is valid for any isolated system.

It follows from this compensation principle that decrease of entropy E and, consequently, increase of order in one part of an isolated system requires an indispensable increase of entropy E and, consequently, increase of disorder in another part of this system for them to compensate each other. Thus, if two (A_1 and A_2) or more open systems interact with each other and together constitute an isolated (closed) system A_3, then the general closed system A_3 remains equilibrium if the change of entropy of one system will be equal to the change of entropy of another system with an opposite sign. The entropy compensation principle as a universal principle for isolated systems must be spread on the family as well if the family is conventionally considered as an isolated system consisting of subsystems: parents and children.

Therefore, decrease of family entropy at the expense of such geniuses and important and influential people in the world politics, culture, and science, as were Pushkin, Lermontov, Dostoyevsky, Mozart, Beethoven, Alexander Macedonian, Napoleon Bonaparte, Lenin, etc. (about 500 people were studied) due to the entropy compensation principle caused the increase of their children's entropy, and as a result their children's creative potential and professional abilities were less important and influential in the world culture and, as a rule, even lower than the mean level (as a famous saying runs, "Nature takes a rest on talented people's children"). Therefore, evidently, natural or artificial craving of some members of the family for making the top in something and for creating prosperity causes the decrease of other members' potential and does not make the whole family happy. Some people disagree with it, they may bring an example when both father and son are academicians. But we speak of people who greatly influenced the world culture, events and processes and not about ordinary academicians. Besides, since it is regularity and not a law, it admits of some exceptions.

The entropy compensation principle is valid only for processes taking place inside the system under study, if it exists inside an isolated system and the influence of the environment or other outer systems comes to one parameter – the degree of the openness of the system.

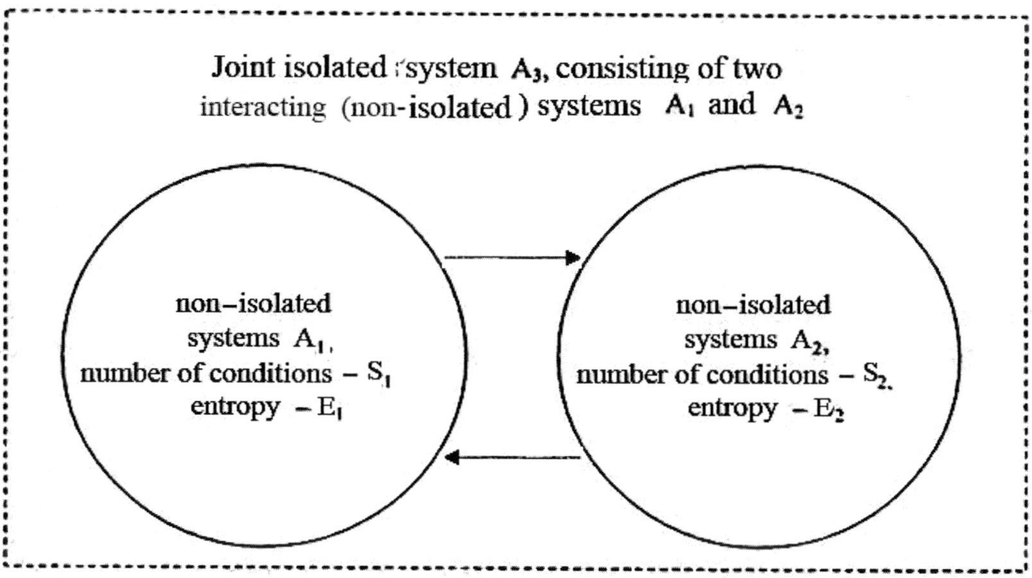

Figure 1.2. A joint (closed) or isolated (not interacting with others) system A_3 consisting of two non-isolated or open interacting systems A_1 and A_2.

It follows from the entropy compensation principle that the USA and Europe are decreasing their entropy (and consequently are improving their situation) at the expense of increasing (as much as or a little more) the entropy of the third countries including Russia and the Commonwealth of Independent States, conquering their product markets, hampering the development of their industry and agriculture, expanding the role of dollar and discharging remains of entropy to these countries. The International Monetary Fund (IMF) also helps to decrease entropy of the developed countries at the expense of increasing entropy of the third countries, Russia and the Commonwealth of Independent Countries by controlling their reforms, their finances and economies as a whole in order to finally weaken their competitiveness, interfering with their entering the global market and conquering developed countries' product markets, preventing them from discharging their waste of entropy to these countries. On the other hand, the developed countries supply the third countries with modern technologies, for example, mobile telephones and other information-communication means, which improves these countries' functioning and thus decreases their entropy. But it is not sufficient, and the deterioration caused by some other reasons is not compensated for.

Such actions ensure sustained development of the USA and Europe. However gradual market exhaustion for their traditional products and decreasing expansion of the dollar in these countries may cause a crisis in the US economy if a new information sector of economy producing new information technologies does not ensure the preservation and extension of the market (new computers, programs, networking, etc.).

Successful development of one group of people or one stratum of society, of one family or one country and, consequently, decrease of entropy is frequently possible only under simultaneous increase of entropy in other strata of society, in other families, countries and environment [11, 25].

Therefore, as a rule, it is impossible to achieve global progress and welfare. However, to improve the situation, as it will be shown below, it is necessary to learn to control the system and to transfer all the excessive entropy (ΔE) to other systems, to the environment or to outer space [11].

The world is impossible without wars, weapons, conflicts, natural calamities and catastrophes. However, as it follows from the entropy compensation and entropy fluctuation principles, it is possible to establish such a control around the point of entropic equilibrium that the decrease of the amplitude and frequency of entropy fluctuation will cause less trouble because it will happen less frequently and its range will be lesser.

As it will be shown below, efficient ways of the system development may be determined according to the dynamic entropy equilibrium between order and disorder.

Any human activity ought to be directed so that to accelerate the decrease of entropy in comparison with its increase in the system.

It is known that two opposite processes are simultaneously taking place in systems. They are the processes of changing entropy (E) and negentropy (NE), where NE represents a quantitative degree of order and organization of the internal structure of the system. Entropy and negentropy act in opposite directions and a negentropy increase in the system causes the same decrease of entropy. However, entropy and negentropy in the system change according to their own regularities [11].

Together with the increase of entropy in a system its indeterminacy increases as well. And in order to decrease this indeterminacy of the system negentropy (information, knowledge, and order) must be introduced into it. If the increase of negentropy in the system leaves behind the increase of entropy, then progressive development and organization prevail in the system and vica versa if the increase of negentropy lags behind the increase of entropy, then destructive processes predominate in the system and disorder grows in it.

Therefore, the global purpose of society must be ensuring accelerated or at least equal increase of negentropy as compared with the increase of entropy in the system. Thus, for instance, if in the future the mankind succeeds to get energy people need not from oil, coal or gas, but from the gravitational field or magnetic field of the Earth, or else from hot or cold thermonuclear fusion, then the entropy of the planet will not increase. It is significant to learn to control entropy-free or low-entropy processes for obtaining necessary energy from the gravitational field, thermonuclear fusion, the sun, wind, etc. which after being used will not increase the entropy of the planet.

The main reasons for increasing entropy in the modern world are as follows [1]:

- criminality, terrorism, destructive wars, ethnic, religious, international conflicts;
- natural and technogeneous disasters and catastrophes, exhaustion of energy and natural resources;
- epidemics, diseases, quick ageing, decrease in population;
- economic and social crises;
- lack of confidence in authorities and mass media;
- fears for the future;
- enrichment not according labor, corruption and undermined discipline;
- injustice, indeterminacy and despair.

For decreasing entropy it is, evidently, necessary first of all to increase the negentropy of public conscience and through it to increase the negentropy of the material world.

1.3. Regularities of Entropic Dynamic Equilibrium or Balance

One of the 28 universal and general-system regularities considered in [22] is the regularity of entropic equilibrium or increase and decrease of entropy (or negentropy) in an open system [8, 22, 32, 33]. This regularity determines the condition of dynamic equilibrium between order and disorder, between organization and disorganization of different systems and to a great degree predetermines the reasons for conflicts in society and family, wars, natural disasters and ecological catastrophes, etc. Entropic equilibrium between order and disorder in a system determines the processes in space and on the earth and living conditions and development perspectives on our planet.

It is a universal regularity, it operates in any system in which it is acceptable to use a conception of probability of events with which the very definition of entropy is connected. In our everyday life oversaturated with technical equipment the probability of technogeneous accidents and catastrophes is constantly increasing. In experts' opinion, by 2003-2005 the main existent resources including technological equipment will reach 80-90 per cent of wear, and the lack of preventive repair and renewal will cause a great number of accidents and catastrophes. Therefore, it is necessary to find out high-risk zones, where the probability of destruction and degradation is very high so that to artificially decrease entropy in such zones.

The world around us consists of open systems, whose degree of openness (α) lies within $0 < \alpha < \alpha_{\max}$ and depends on the intensity of energy (En), information (I), and substance (G) exchange with the environment or other systems. $\alpha = 0$ are possessed by a completely isolated (closed) system where no energy, information and substance exchange with other systems takes place. All the processes in such an isolated or closed system are accompanied only by the increase of entropy of the system (fig.1.3), $\alpha = \alpha_{\max}$ must have nonreal completely open (absolutely non-isolated) systems in which all the processes are accompanied only by the decrease of entropy (Figure 1.4) [8, 22, 32, 33]. However there are no absolutely closed or open systems in nature.

In real open systems, in which $0 < \alpha < \alpha_{\max}$, all the processes are accompanied by both increase and decrease of entropy of the system (Figure 1.5). Such systems oppose and compensate each other. And thus the final sign, changing total entropy E or negentropy NE towards either increase of entropy (sign plus, $\Delta E > 0$) or decrease of entropy (sign minus, $\Delta E < 0$), is determined [8, 22, 32, 33]. As a result we get either more order and organization, or more disorder and disorganization.

At the point where the actions of both opposing entropy signs completely compensate each other entropic equilibrium (E_e) or a critical level of the system organization (E_c) is established. The point of entropic equilibrium between order and disorder, between organization and disorganization in nature is shown in figure1.5. Since at the point of entropic equilibrium (E_e) or at the critical level of the system organization (E_c) the processes of order (organization) and disorganization balance each other – the system becomes stable [22, 23, 32].

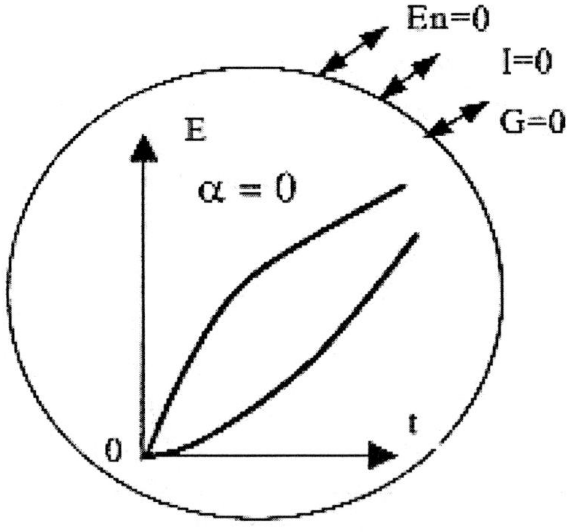

Figure 1.3. Completely closed or isolated system. Entropy increases over time.

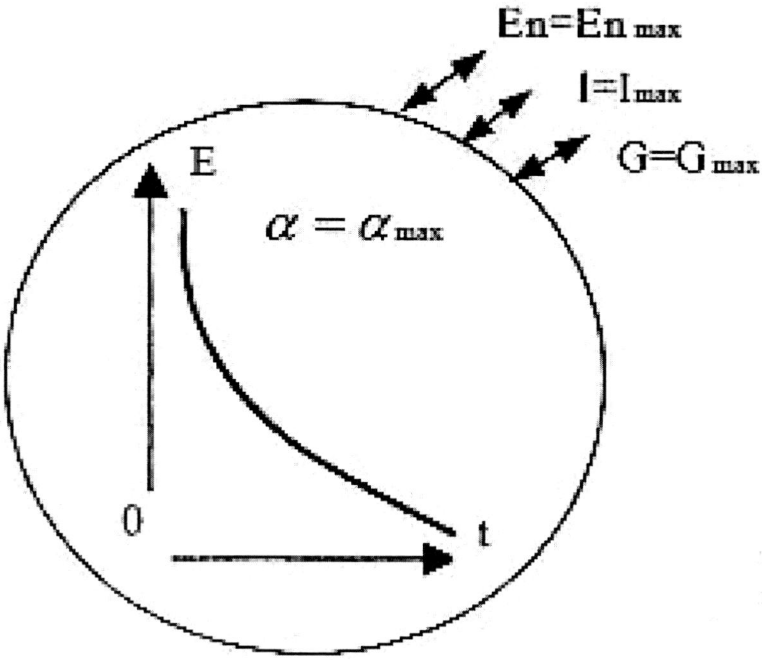

Figure 1.4. Completely open system. Entropy decreases.

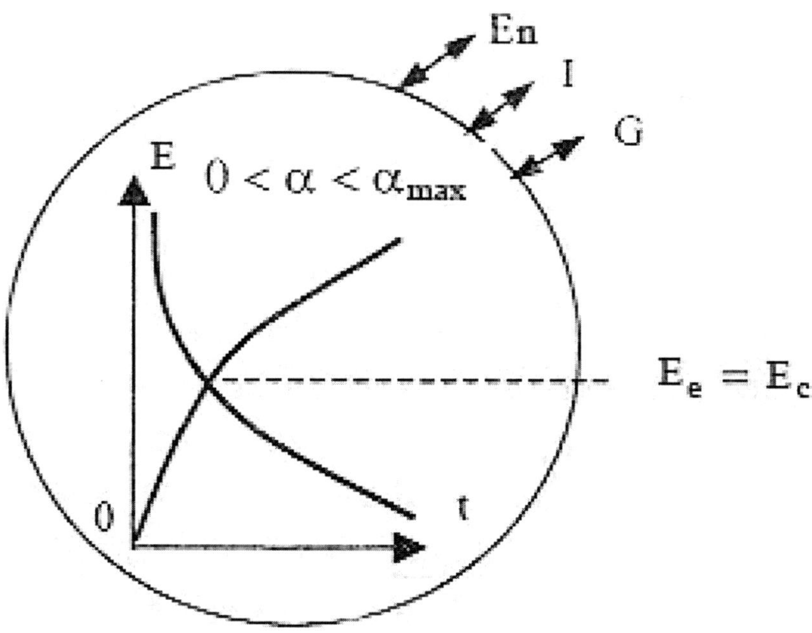

Figure 1.5. Real system with $0 < \alpha < \alpha_{\max}$ degree of openness. Both curves of increase and decrease of entropy are present in the system. The point of their intersection defines the point of entropic equilibrium (E_e).

Figure1.6. shows the first method of controlling entropic processes in the system by means of shifting the line of entropic equilibrium (E_{e1}, E_{e2}, E_{e3})) in the system.

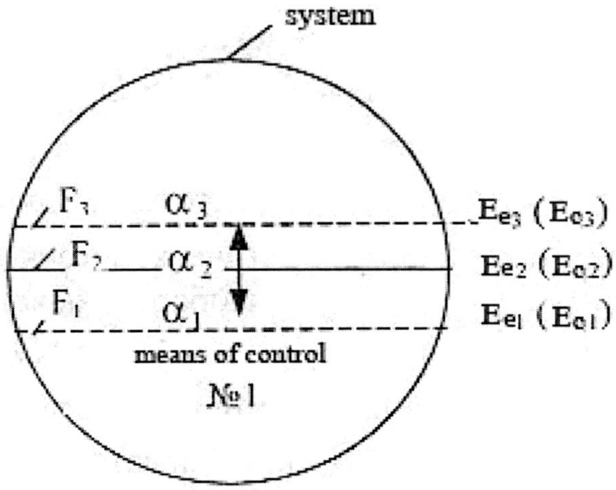

Figure 1.6. Shifting the line of entropic equilibrium $E_e(E_{e1}, E_{e2}, E_{e3}) = E_c$ by means of changing the level of openness of the systems ($\alpha_1, \alpha_2, \alpha_3$) at the expense of changing the external influence (F_1, F_2, F_3).

Indeed, the degree of openness of the system (α) changes uniquely as the external influence on the system changes. As the external influence (F) increases, the degree of openness (α) increases as well, and as the influence decreases, the degree of openness decreases too (Figure 1.6)

The degree of openness and, consequently the external influence (F) uniquely determines the value of entropic equilibrium (E_e) or the critical level of the system organization (E_c) (Figure 1.6). In figure 1.6 three different levels of the openness of the system ($\alpha_1, \alpha_2, \alpha_3$) are shown. They correspond to three levels of external influence (F_1, F_2, F_3) and respectively to three levels of entropic equilibrium or critical points of the system organization ($E_{e1} = E_{c1}, E_{e2} = E_{c2}$ and $E_{e3} = E_{c3}$). The shift of the line of entropic equilibrium or critical point at the expense of the change of the system openness, as it is shown in figure 1.6., is conventionally called method of control No 1.

If a certain concrete system is originally placed (organized) higher than the critical level, i.e. its original entropy (E_{01}) is greater $E_e = E_c$ ($E_{01} > E_e = E_c$) (curve 1 in Figure 1.7a), then in such a system processes of decreasing entropy down to $E_e = E_c$ will prevail and, consequently, the processes of organizing (ordering) the system up to the critical point or the point of entropic equilibrium. But if, vica versa, the original value of the system entropy is lower than the critical level ($E_{02} < E_e = E_c$) (curve 2, Figure 1.7a), then the process of increasing entropy from E_{02} to $E_e = E_c$ will prevail in the system and, consequently, the processes of disorganization will also prevail in the system.

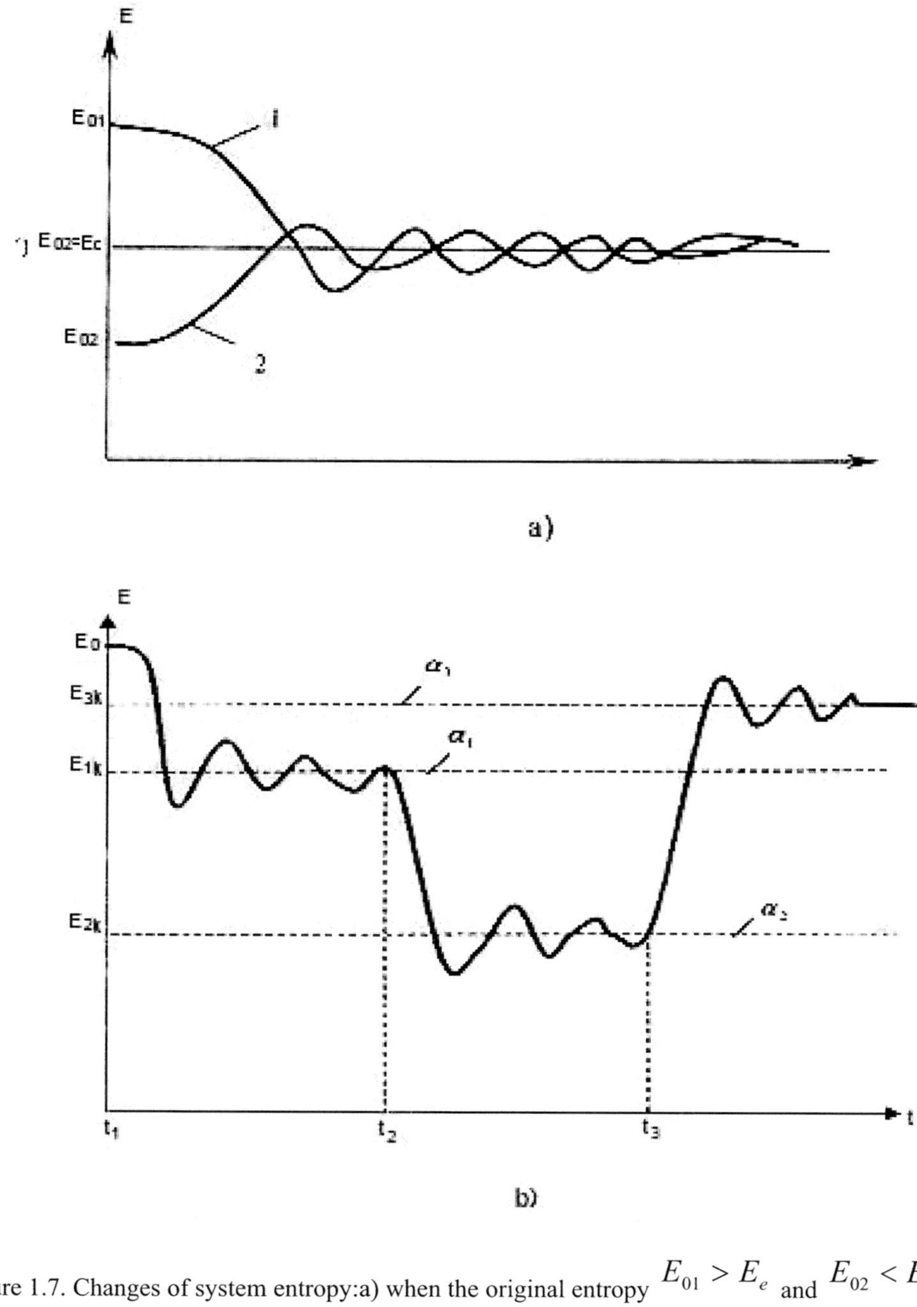

Figure 1.7. Changes of system entropy:a) when the original entropy $E_{01} > E_e$ and $E_{02} < E_e$; b) when entropy ($E_0, E_{1c}, E_{2c}, E_{3c}$) changes its degree of openness ($\alpha_1, \alpha_2, \alpha_3$). E_{1c}, E_{3c}, E_{3c} - values of entropy of critical levels of organization, corresponding to different degrees of system openness ($\alpha_1, \alpha_2, \alpha_3$). The increase of the openness degree at the moment t_2 causes the organization of the system to the critical level E_{2c}, and its decrease at the moment t_3 causes the disorganization of the system to the new critical level E_{3c}.

Because of the system inertia in the process of changing entropy in the system there appear entropic fluctuations relative to the entropic equilibrium or critical level ($E_e = E_c$), which may subside over time and the system will be stabilized. In figure 1.7b the change of the system entropy is shown as its openness changes ($\alpha_1, \alpha_2, \alpha_3$).

The increase of amplitude and frequency of fluctuations reinforces negative processes in nature and society. It reinforces the probability of natural disasters, destruction of ecosystems, increases the frequency and acuteness of accidents, epidemics, conflicts in society, etc.

It may be assumed, that global cataclysms which happened in different historical epochs owe their cyclic character, first of all, to the phenomenon of entropic fluctuations [32, 33].

While transforming the world, man is changing its order and accordingly increases or decreases entropy.

While in times of peace the mankind is busy improving nature in its own way, it unavoidably is approaching the critical level or entropic equilibrium and under its own inertia is aspiring to exceed this level and, consequently, is approaching conflicts, wars, or such a natural disaster, whose scale may be compared with a war. We can tell the beginning of this process by the sharpening of the ecological crisis, a considerable increase in the number of accidents and technogeneous catastrophes, social conflicts, local wars, and any other events which generate a destructive tendency on the planet [32, 33].

When the mankind, busy with improving nature, exceeds the level of entropic equilibrium processes of disorganization will prevail on the earth despite any peace initiatives and ecological programs. But if the openness of the planet is increased, for instance, by means of gradual and large-scale space exploration, the critical level of the organization of the planet or the line of entropic equilibrium will grow considerably which would cause the process of self-organization and only then ecological programs would work.

It should be mentioned that the increase of the level of the openness of the system (α) causes the decrease of entropy and the establishment of order not in all the parameters of the system but only in some of them and, on the contrary, the increase of entropy in other parameters and disorder of the system in them. Thus, for example, the increase of the openness and transparency of international borders and decrease of customs and other llimitations improve labor, cultural, trade, economic and other characteristics, and in these parameters the entropy of the system and, consequently, disorder decreases, but in other characteristics of the system, for example, such as the influx of refugees, freer movement of drugs, terrorists, etc., entropy and disorder increase. Thus, increase or decrease of entropy never occurs in all the parameters of the system and, consequently, a general transition from disorder to order does not occur in all the parameters of the system.

If, for instance, the environment is deliberately aggressive in certain parameters and the openness of the system increases, then such an aggressive environment may degrade and even destruct the system. Therefore, the system should be protected from the aggressive environment by means of decreasing the degree of openness, for instance, efforts should be taken to limit access from abroad for terrorists, drugs, pornographic cassettes, low-quality products, but, on the otherhand, to throw open the market door to local commodities and commodities from the Commonwealth of Independent States. If the environment is not aggressive, then the country which on its borders limits its openness will experience the strengthening of destructive processes in the sphere of economy, culture and entropy will increase in these directions. When because of high custom duties and high prices the supply of foreign quality products at our markets is reduced, our dishonorable manufacturers start producing analogous, but low-quality products passing them for foreign commodities and thus increase entropy and disorder in these parameters [22, 33].

Entropic equilibrium or entropic balance exists only at the critical level of the system organization, when disorder and disorganization balance order and organization, and the system restores its stability. Any deviation from the equilibrium point in any direction causes either excessive order and organization or excessive disorder and disorganization. Each system, both natural and social, strives for entropic equilibrium ($E_e = E_c$), to the equilibrium between order and disorder.

Thus, on the planet just like in a system after a long-term perturbation and restructuring delicate equilibrium has been restored between order and disorder, which compensate each other at the expense of interaction between the Earth and Cosmos. During the period when there was no entropic equilibrium on the planet, new mountains, lakes, seas, oceans were in the process of forming and mountains and seas were growing and reducing in their size. Nowadays these mountains and seas, oceans (Figure 1.8) almost never change in size, which points to the fact that entropic equilibrium and a certain stability have been restored [22, 33].

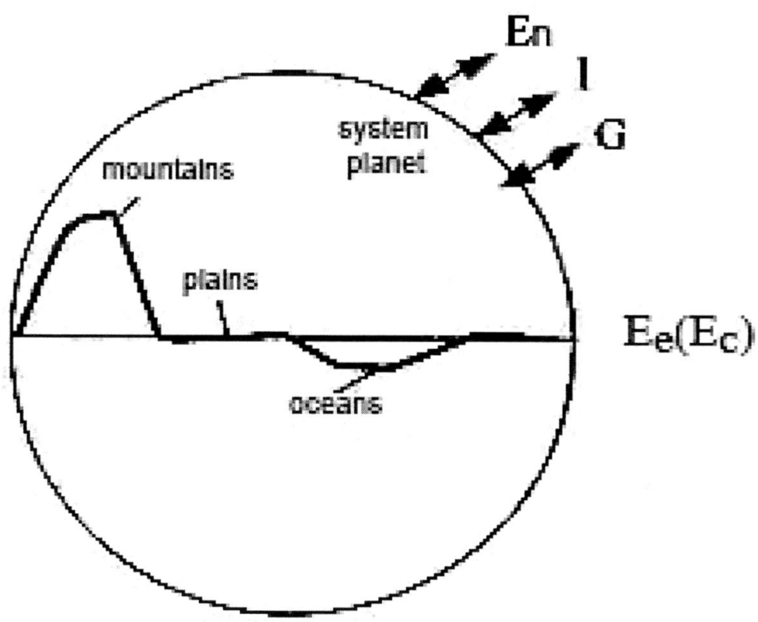

Figure 1.8. There exists a delicate entropic equilibrium on the planet between order and disorder.

When the mankind upsets the entropic equilibrium of the planet and realizes excessive by value $\Delta E < 0$ order and organization at the expense of building additional towns, plants, electric stations, dams, etc., the entropic equilibrium is upset by value ΔE, overstepping through inertia the critical line of equilibrium ($E_e = E_c$) (Figure 1.9).

However, as nature cannot bear the disturbance of equilibrium, it always aspires to recover the entropic equilibrium between order and disorder whenever it is upset. Therefore, nature must do away with excessive order ($\Delta E < 0$) which occurs at the other end of the equilibrium and which the humankind creates through inertia. For prompt elimination of excessive order ($\Delta E < 0$) and recovery of equilibrium wars and natural disasters may break out, conflicts may erupt.

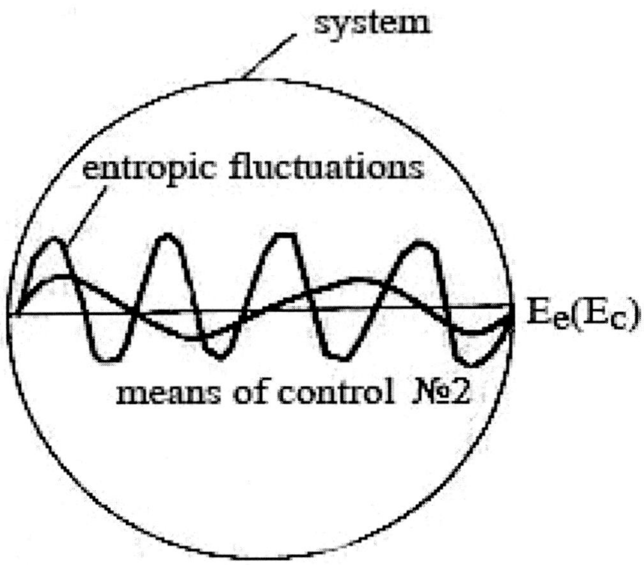

Figure 1.9. Entropic fluctuations relative to $E_e = E_c$ and control of amplitude and frequency of entropic fluctuations.

However, nature by means of wars, natural disasters and conflicts through inertia destroys more than is necessary for recovering equilibrium and thus goes over to the other side of the equilibrium line $\Delta E > 0$, which causes greater destruction and disorganization than it was necessary for the recovery of natural equilibrium. Then again the humankind starts eliminating the excessive destruction caused by wars, natural disasters and conflicts. But in the process of recovering equilibrium through inertia the mankind will again cross the line of equilibrium and so on and so forth. Therefore, entropic fluctuations emerge, as it is shown in figure 1.9, around the equilibrium line. From the picture of entropy fluctuations it follows that it is utopian to create such a society on the earth where there would not be any wars, conflicts or natural disasters and catastrophes. It is impossible unless the humankind learns to control the amplitude and frequency of entropic fluctuations in order to reduce them and, consequently, to make calamities less disastrous and avoid them more frequently. This second method of controlling entropic processes is in a simplified way shown in figures 1.9 and 1.10. Obviously, no ecological programs can be realized unless they are tested for entropic equilibrium.

The third method of controlling entropic process in the system is shown in figure 1.11. Here for ensuring better order and better organization in the system it is necessary to “take” excessive (remains of) entropy out of the system to the environment, to the universe or to other systems [11, 22].

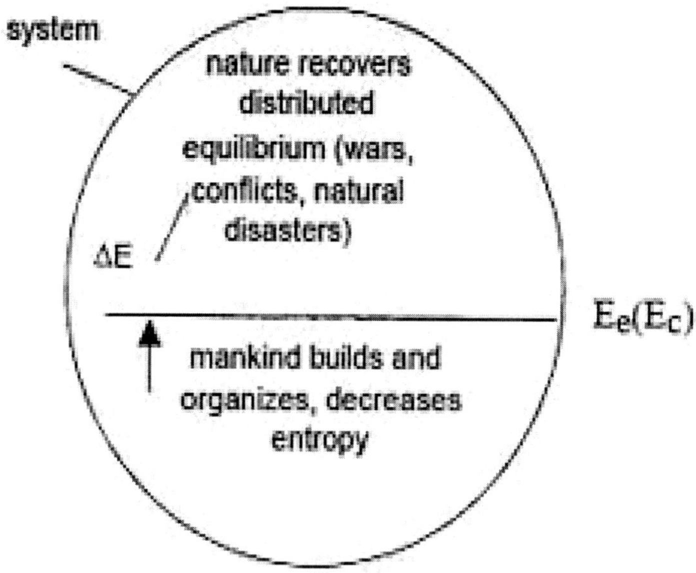

Figure 1.10. Inertial interaction of man and nature (wars, natural disasters and conflicts in society caused by the upset equilibrium).

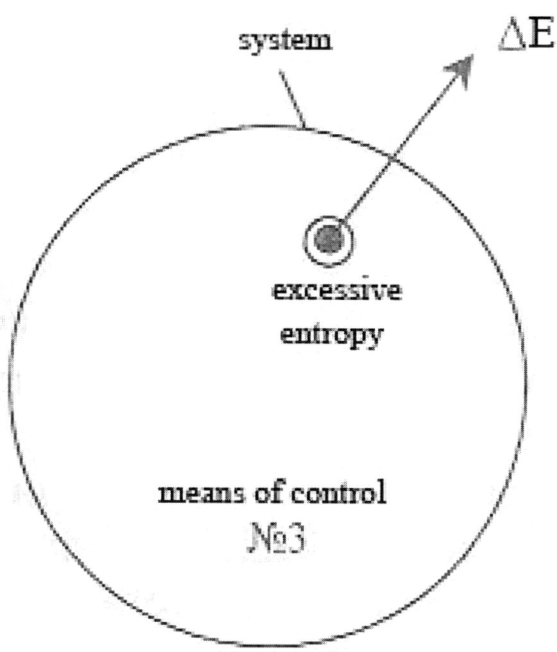

Figure 1.11. Taking excessive entropy out to the environment, to other systems or to the universe.

It should be noted that in order to make a considerable opening of the planet it is necessary to develop space on a large scale (the Moon, the Mars, etc.). Only under a large-scale development of space the process of self-organization of the planet will prevail and the probability of destructive events will be reduced. A question arises concerning the ways of doing it in practice.

The development of the Moon and Mars can be implemented only by joint efforts of many countries. Consequently, instead of wars and conflicts many countries will have to create a joint program for developing the Moon. Then the Mars and other planets will be developed. And each time the mankind delays the next in turn large-scale opening of the planet, a menace of universal destruction will arise. In other words, the planet, we live on, continuously needs opening. Therefore, it is necessary to make joint efforts with many other countries in order to implement a common program on developing the Moon, then after completing it, to open our planet by joining in with other countries in investigating the Mars and then other planets. Thus, to avoid violent conflicts we must always open the planet and change its critical level. And with this purpose in mind the leading countries must always unite their efforts in investigating different space objects.

It does not mean that the development of science and technology, culture, economy and improvement of welfare will decrease the general entropy of the system in all its characteristics. In fact, while in some parameters of the system entropy decreases, it increases in other parameters, since the population growth on the planet will intensify competition and increase waste products of human activities. The population growth will cause new conflicts, wars, ecological problems, etc. [11, 22].

A lot of practical examples around us demonstrate the operation of the regularities of entropic equilibrium, or increase and decrease of entropy in an open system. Let us consider some of these examples.

An obvious case of the given regularity is illustrated by inoculation with attenuated viruses which forces the human organism to organize itself to such a critical level of the system organization (E_c) which is sufficient to resist a real harmful virus and not to fall ill. However, some time later the critical level (E_c) will change and the entropic equilibrium will be upset, as a result the organism will become somewhat disorganized and again the organism will be subject to diseases [22, 32].

Another example of the given regularity is observed in the impact of the abatement and enhancement of physical exercise on a person. If you do different muscle-strengthening exercises, i.e. experience external influence, you can keep fit and maintain the process of self-organization in your organism [22, 33].

One more example is connected with educating a child. School, parents and different hobby groups actively influence the child by educating him/her. They improve his/her knowledge, culture, physical and mental condition and, consequently, increase the level of his/her openness as a system: the child more actively exchanges information with the social environment and thus decreases his/her entropy. If parents and school for some reasons temporarily neglect the educational process, it weakens the external influence on the child as a system and thus the level of his/her openness (α) to information, energy and substance decreases. Because of the decrease of the openness the process of degradation will prevail in the system (child). The degradation manifests itself in the deterioration of the child's physical and mental condition, his/her knowledge and cultural level [22,33].

And now we will say just a few words about positive biological entropy and the origin of "entropic disease". It is known that in addition to information a modern person has to endure a flow of positive biological entropy. Different diseases and first of all diseases connected with ageing or cell change and with the loss of their organization and their transformation into an avalanche-like multiplying biological forms (malignant tumor) – all these cases represent different forms of biological entropy. More than 59% of people die of similar "entropic diseases" in different countries. One of the main reasons for the increase of biological entropy is viruses which ruin DNA and RNA and substitute them by penetrating into cells. Exactly those cells which have increased entropic information are subject to enhanced uncontrollable multiplication [8].

A mathematical model of demographic growth in the world was offered by professor S.P.Kapitsa and other scientists. This model conforms to the data on changes in population beginning with the 18-th century and shows that in the 21-t century the population of the planet may reach an extremely great number. As the planet will not be able to endure such overpopulation and entropic equilibrium in the nature will be upset, the nature will by all means "take certain measures" for reducing the population and recovering entropic equilibrium. This may happen, for example, due to a mass epidemic, AIDS, immune or other diseases, wars, catastrophes, etc.

It should be mentioned that one of the most important missions of the humankind on the planet is struggle against entropy, therefore, each person must fulfill an antientropic or negentropic function. Morality and ethics are considered to be negentropy. They oppose disorder and chaos, i.e. maintain order and life against chaos and disorder which are characteristic of entropy.

From the standpoint of the regularities determining the increase and decrease of entropy in real open systems the doctrine of noosphere seems utopian, especially its cosmic version which considers noosphere as a complete annihilation of evil, the common good without wars and weapons. Indispensable entropic fluctuations around the dynamic equilibrium indicate that there always will be wars, conflicts, natural disasters, environmental catastrophes, etc. The main task of the mankind is to reduce the amplitude and frequency of fluctuations by rational behavior and control and thus to reduce the frequency of calamities and mitigate their results.

Epochs follow one another, politicians come and disappear from the scene, new ideologies originate and collapse, means of aggression and defense are improved, but the world still lives like in a vicious circle for the humankind is still unable to achieve conflict-free existence which is explained by entropic fluctuations around the line of entropic equilibrium. New conflicts erupt instead of the old ones in which the world powers may be involved. Thus, the American journal "Newsweek" in 2002 published a forecast of probable conflicts on the planet till 2012. As a new destabilization weapon it is offered to use powerful directed electromagnet impulses coming from low-orbiting spaceships that can paralyze bank computer nets, transport means, air, ground and sea means of communication as well as the Internet and thus paralyze all the vital activities of a country.

The necessity of keeping certain entropic dynamic equilibrium in any natural and social systems shows that nature does not bear the disturbance of entropic equilibrium and balance between order and disorder, between organization and disorganization and if entropic equilibrium is disturbed, different measures are taken by nature or society to recover it. From the conditions necessary for maintaining necessary entropic equilibrium it follows that it is impossible to ensure good life for all the countries and all the members of the world community, it can be achieved only by minority groups of society. Therefore, in a situation when there are limited natural and food resources on the planet the aspiration of the underdeveloped countries and peoples to increase their consumption level to that of the developed countries is utopian and cannot be achieved. It means that the achievements of some countries and peoples or of a certain stratum of society in raising their living standard is possible only if at the same time the living standard of other countries and peoples, or of another stratum of society decreases. Therefore, universal progress and welfare, including communism, are impossible, otherwise entropic equilibrium in nature and society will be upset for a long time which is impossible. Proceeding from the volume of natural and food resources existent on the planet today and from the regularities of maintaining entropic equilibrium in nature and society the following two scenarios of social-economic development of the world community are possible:

- the first scenario – high consumption and cultural levels and prosperity are guaranteed to the "gold billion" of the population of the developed countries and subsistence minimum corresponding to survival minimum is guaranteed to the majority (6-7 billions) of the population of the underdeveloped and developing countries;
- the second scenario – the same moderate consumption level is guaranteed to all the countries and all the members of the world community. In this case everybody lives satisfactorily from the point of view of consumption level (as under socialism), but spiritually much better, and the regularity of entropic equilibrium is not upset. Therefore, the system is stable.

It should be noted once more that until the mankind learns to efficiently apply the three methods of controlling entropy processes in systems including the critical level of the system organization (Figure 1.6. – the first control method), amplitude and frequency of entropic fluctuations (Figure 1.9 – the second control method) and take excessive entropy out to the environment or space (Figure 1.11. – the third control method) for stabilizing the system, it will be necessary to constantly increase the degree of the openness of systems by means of external control influence [22, 32, 33].

Next we will briefly consider how interaction with the environment influences the change of entropy in open systems if we introduce the concept of entropy stated in [22, 23]. To do so, we will designate by $E(X_0)$ the entropy of the condition of the system when it experiences no external action from the environment, where X_0 - a variable that describes the condition of the system in absence of external action. As a result of external influence the condition of the system will change and, consequently, the system entropy corresponding to this condition will change as well. This is called conventional entropy E(X_0/X_1) where X_1 - a new variable describing a new condition of the system which results from external influence from the environment [22, 23]. It is obvious that conventional entropy $E(X_0/X_1) \leq E(X_0)$ because thanks to external influence the system increases the degree of its openness (α) and, if the environment is not aggressive, if there are no drugs and terrorists, the system decreases its entropy in non-aggressive parameters. If the external influence on this system is enhanced, the system will open even more and the conventional entropy of the new state will decrease even more considerably $E(X_0/X_1X_2) < E(X_0/X_1)$ where X_2 is a new variable which describes a new change in the same system.

Therefore, we will have $E(X_0) > E(X_0 / X_1) > E(X_0 / X_1 X_2)$ and will in a general case receive:

$$E(X_0 / X_1 X_2 ... X_c) < E(X_0 / X_1 X_2 ... X_{c-1}).$$

It is true that if the external environment is aggressive and comprises, for example, drugs, terrorists, diseases, AIDS, etc., the result may be the opposite: $E(X_0 / X_1 X_2 ... X_c) > E(X_0 / X_1 X_2 ... X_{c-1})$.

Each degree of openness (α) of the system has its own respective stationary value of entropy $E_c = E_e$ determining the equilibrium. Thus, if $E_\alpha > E_c$, the processes of entropy decrease and system degradation will prevail until entropy equilibrium is recovered.

And at last it should be noted once more that equilibrium between order and disorder as a whole in all the parameters of the system implies their inequality for separate parts and separate parameters. A complex change of the borders between two equal opposites – order and disorder – corresponds to the evolution of nature. The increase of order, organization of the system is caused by the increase of order in certain parameters of the system and by the increase of disorder (disorganization) in other parameters, and not by general transition from disorder to order or vica versa in all the parameters of the system. And the stability of the system is determined by the correlation of the values of order and disorder degrees for corresponding parameters according to Fibbonacci and the "gold proportion" method [22, 28, 32, 33]. It is shown in these works that entropy equals to the quantitative degree of disorder only if equal probability of events is postulated. As for a general case, if we deal with unequal probability of events, entropy equals to the sum of degrees of order and disorder. The modern equilibrium model considers cycle of nature where the correlation of parts and the whole is described according to the "gold proportion" rule; it is sometimes called a harmony of nature law which is observed when a number of general features of living nature and abiocoen, society and economy are described. The equality of degrees of order and disorder determines the equilibrium of the cycle of nature and society. If the "gold proportion" rule for the degree of order and disorder is realized in three parameters: a) changing the structure of the elements; b) changing the interaction between the elements; c) moving in space, and a couple of parameters are described according to Fibbonacci method, the system will keep equilibrium and stability [28, 32, 33].

Some scientists think that the "gold proportion" dividing the whole into two unequal parts of 2/3 and 1/3 of the whole is the expression and creation of divine proportions and the geometrical relation according to the "gold proportion" (2/3=0.618 and 1/3=0.382) resonates with acts of nature.

It is ascertained that the "gold proportion" phenomenon comprises not only fundamental proportions, but basal metrics as well which is able to harmonize many technical, economic, financial, social and other relations.

The "gold proportion" phenomenon is connected with brain operations with information; and human intuition (the reality of intuitive forms) has such a structure which induces a person involuntarily to isolate two unequal parts (portions) – 0.618 and 0.382 which are correlated between each other by means of the "gold proportion" and whose sum equals to one. The "gold proportion" in nature has always been considered to be a symbol of wisdom, harmony, *perfection*, beauty and exists in ancient and modern architecture, painting. culture, management psychology, finances, economics and technology, beautiful nature.

It was found that during a reflexive synthesis of the best imaginable alternative decisions the human intellect is commensurate with the harmonious structure of the "gold proportion".

1.4. The Oscillatory and Cyclic Character of System Functioning

The oscillatory and cyclic character of functioning of different systems is the next important general system regularity that in many aspects determines the development of states, regions, civilizations, international, and other relations. Such regular oscillatory and cyclic process of the system accompanied be continuous transmission from one condition to the opposite one and back again is happening non-stop. If we analyze the history of the development of states, we will see two opposite processes: the first one is the tendency towards uniting in one strong state under single control with a good economy and the other one is the tendency towards breaking up into separate areas, regions, towns, weakening of economic strength and military might and centralized power of the state and craving for independence, separatism. On the one hand, it means centralization, territorial integrity and, on the other hand, it implies disintegration, decentralization, and intensification of self-detaching process. The analysis of the processes aimed at exposing oscillatory and cyclic character shows that nowadays consolidation (pasting) of separate states and nations greatly promotes the globalization of economy and the existence of transnational corporations requiring free transportation of commodities, raw materials, products, people and absence of strictly fixed state borders. The process of uniting states and creation of common currency, legislation, government, parliament, etc. is already taking place in Europe.

At present national states (ethnos) constituting this or that civilization (superethnos) are gradually intensifying their contacts. National states are supposed to disappear over time since a number of states will together create their own civilizations which occupied the territories of large regions and united people according to their religion, history, cultural level, traditions, mentality, etc. Each civilization may turn into a integrated political unit, and in the future the world will be formed by interaction of seven or eight such major civilizations as, for example, Western, Orthodox, Islamic, Indian, Latin American, African, Japanese and Confucian (Chinese). Today the world structure is already formed like a certain set of civilizations. Regional economic organizations like Western Europe, South-Western Asia, Latin America, etc. are established. The process of economic globalization is underway. It results in wiping out national distinctions, abolishing state borders and creating new civilizations.

Today oscillatory and cyclic character of the historic process represents a certain part of the universe, planet and biosphere with their integration processes in the world (European Unit, UNO, NATO, Roman Club, etc.).

Numerous scientists both in our country and abroad knew that the world develops in cycles and certain processes and phenomena are repeated once in a definite period of time. Russian scientist Chizhevsky studied the cyclic character of solar phenomena and their affect on people and important processes on the earth, scientist Kondratiev (Kondratiev waves) studied the periodical character of economic crises, etc. The study of the oscillation of cyclicity, rhythmicity helps to study the periodicity of crises, the ups and downs.

By way of example, let us consider the influence of oscillatory and cyclic regularity in the history of Russia on the development of Russia, on its integrity, power, economy, military success [22].

Thus the oscillations of the Russia's power are thoroughly studied in [21] over the period of more than 1.200 years and a consistent cyclicity or periodicity of analogous situations in the country is found (Figure 1.12.). For evaluating situations in Russia over the period of time of 1.200 years the following indices are used: territorial and administrative integrity of Russia, its economic development, military success, internal power struggle. As it is seen in figure 1.12, in more than 1.000-year history of Russia (862-1989) alternate changes of situations from a strong and integrated country to a weak and disintegrated one and vica versa are observed in every certain period of time. A.Parkhomenko established distinct oscillatory regularity of Russia's transition from a powerful and integrated state to a weak and disintegrated one, and then again its reversion to a strong, centralized condition. A period of time in the course of which Russian state is successful the scientist conventionally called "peak", and the most unsuccessful periods during which the country is disintegrated, economically weak, has military setbacks, and the population seeks to change the power, is conventionally called "cavity". The periods of time, during which the state moves from its "peak" to its "cavity" and back again, are conventionally called ups and downs. As it is seen from figure 1.12, three recurring cycles are found in Russia's history (the duration of each one equals to 375-376 years). Each of the three 375-376 year cycles comprises four consecutive "peaks" (ups) and four "cavities" (downs) and, besides, four periods of time between ups and downs, when Russia was in the process of transition from a weak disintegrated to a strong integrated state and, vice versa, when it was on its way from a strong integrated state to a weak and disintegrated one [18].

A graphic picture of three typical cycles of Russian history is shown in figure 1.12. This graphic picture covers the period from 862 to 1989. As it follows from the graphic picture, the first cycle started in 862 and ended in 1238, i.e. the cycle lasted for 375 years. The second cycle started in 1238 and ended in 1613 (its duration is also 375 years).

The third cycle started in 1613 and ended in 1989 (it also lasted for 375 years). If we continue this regularity, then by analogy we may consider 1989 (disintegration of the USSR) as the beginning of the fourth cycle which will end in 2364-2365 (1989+375 years).

From 862 to 1989 in the course of three cycles Russia 12 times was united in a strong integrated state and 12 times broke up into separate princedoms and towns, became a weak and disintegrated state having a weak economy and experiencing military setbacks [21, 22].

If we use the graphic picture shown in figure 1.12 as a base for the development of the Russian state, then with a certain degree of probability we can predict (prognosticate) the picture of Russian history in the fourth cycle (1989-2364) applying the results of the first three cycles.

The fourth cycle began in 1989 with the disintegration of the USSR and dramatic decline of Russia. The first recovery of Russia and the achievement of intermediate (stronger) position in comparison with the previous three cycles must start after14-17-32 years from the beginning of the fourth cycle (1989), i.e. in 2003, 2007 or 2021, and the "peak" of the statehood must start after 76-90 years from the beginning of the cycle, i.e. in 2065-2079 years.

By analogy with the previous cycles, Russia within 375 years must go through four periods of ups and downs. And as it is shown in figure 1.12, at the beginning and at the end of 375-year cycles Russia is disintegrated into small princedoms, regions, and towns. It seems that there exists something that controls the country, creates an integrated state from small parts, and then breaks them up into small parts, and everything starts all over again [21].

It is evidently impossible to predict future historic processes with high probability only on the basis of oscillatory and cyclic regularities of the country's development and analogous periods and situations in the previous cycles. However, applying some analogy in historic cycles and knowledge transfer may help by means of specific observation about history to determine the regularities of social evolution in different historic periods [21, 22].

1.5. The Regularity of the System Potential Dependence on the System Structure or the Type of Interaction between Its Structural Elements

1.5.1. The System Potential

The system potential or the resources of the system constitute the basis for its effective operation. Therefore, the development of the system potential or its resources is of paramount importance.

By the potential or resources of the system its essential resources are implied (labor, economic, social, military, political, intellectual, spiritual, etc.), which may be set in motion in order to achieve certain goals of the system [4]. For example, we can speak about military, economic, social, scientific, cultural, intellectual potential of a society, state, family, person as a system.

The potential or performance capability of a system determined by its resources is conventionally divided into four groups of elements, each of which includes both quantitative and qualitative components of corresponding resources: 1) labor; 2) material resources (finances, property, technical equipment); 3) information resources (control systems, technologies, techniques); 4) timing budgets (available time).

It is obvious that the growth of one of the elements of the potential makes it possible to save other resources while carrying out a certain task.

As the system potential consists of qualitative and quantitative characteristics, its growth is possible by means of increasing both qualitative and quantitative elements. It is true that there is some kind of interrelation between the qualitative and quantitative characteristics in each separate type of activities.

The potential of any system determines the resources that are used for performing some tasks and, therefore, they are actualized. But some resources are not used yet, but may be used for performing a system task and, therefore, they are non- actualized.

The problem of system control is simplified or even reduced if the system has a bigger quantitative potential and an opportunity to use it as the need arises and, therefore, at the right moment certain elements "are switched on" and after completing the task they "are switched off". Simultaneously, special measures should be taken for maintaining the non-actualized part of the potential in operating condition [4].

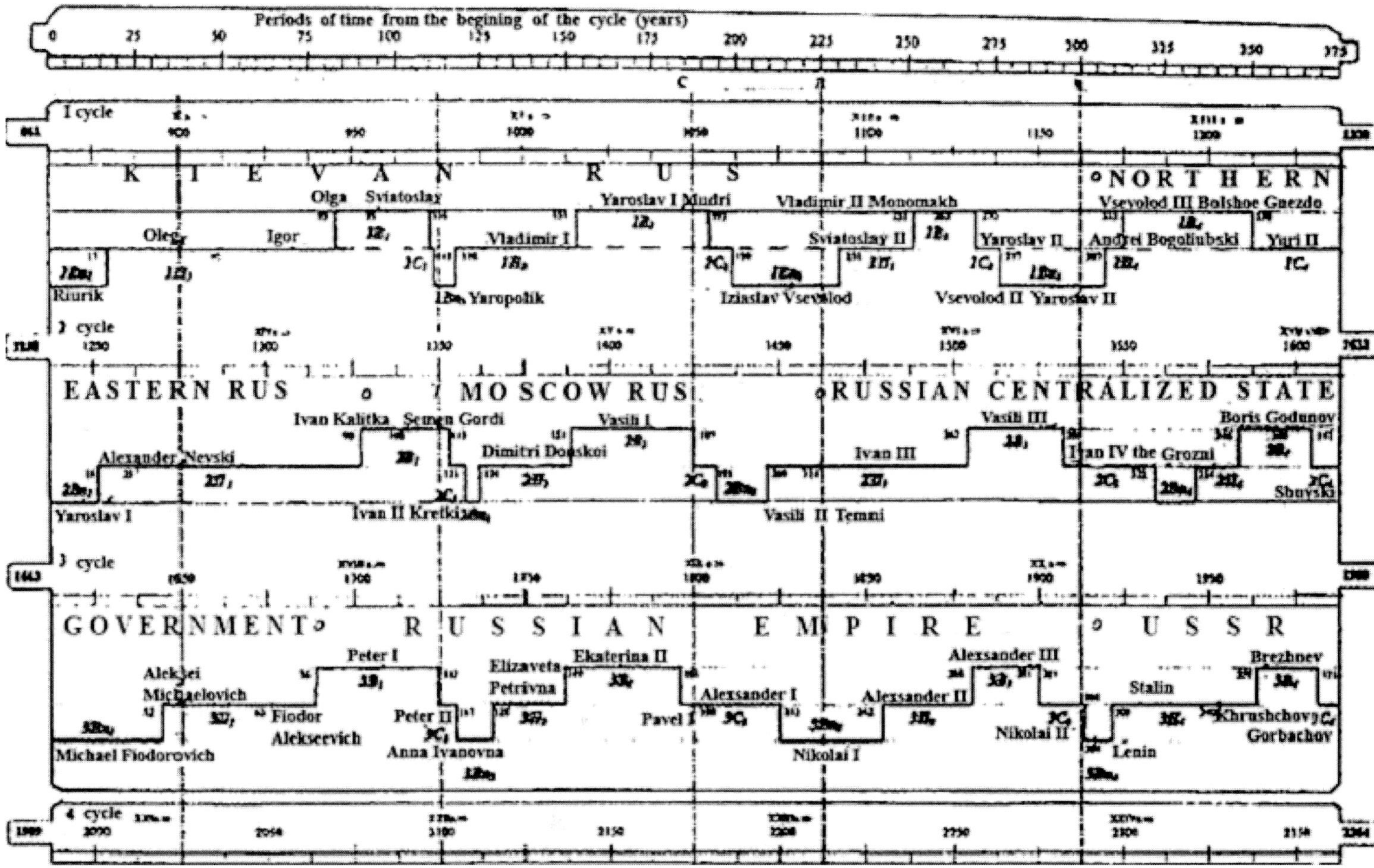

Figure 1.12. Periods of Russia’s power.

The system puts non-actualized potential into operation for ensuring effective performance of the assigned task. Figure1.13 shows a diagram using both actualized and non-actualized elements of the system potential. Both actualized and non-actualized parts of the system potential are used for effective fulfillment of the assigned task.

Social systems (just like other systems) develop in cycles and experience periods of both pickups and downturns (crises) in their history. Herewith, a certain surplus of resources that was received during the period of pickup should be saved in non-actualized type of resources to be used during periods of downturns and crises. The accumulation of resource provisions during the period of pickup makes it possible to neutralize oscillation results at the bottom points of the cycle. Besides, the accumulation of resource provisions is necessary for overcoming unforeseen circumstances.

Maneuvering the potential of systems and its effective use is ensured due to the possibility to concentrate means and resources in separate areas of system operation on a tight timetable.

Some system resources can be put into operation on a tight timetable (quickly) and other resources (elements), vice versa, cannot be prepared for service on a tight timetable.

It follows from the above-mentioned that the potential of the system can be destroyed comparatively quickly, and the recovery of the system potential takes a comparatively long time.

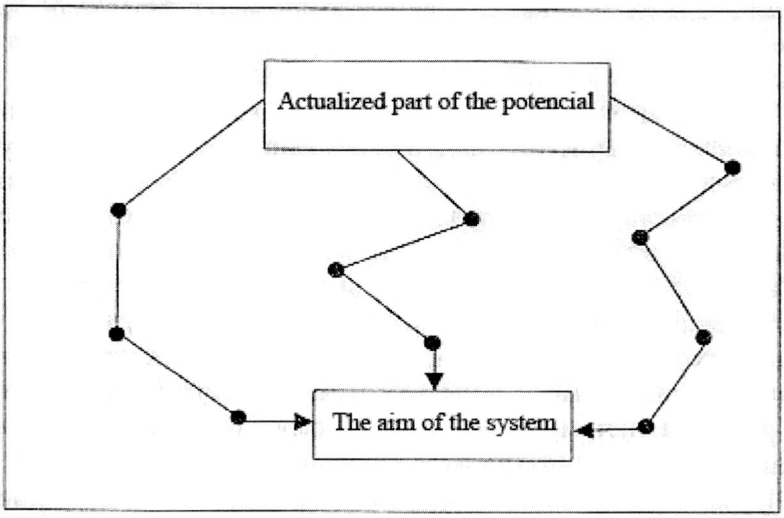

Figure 1.13. Simplified diagram of using actualized and non-actualized parts of the social system potential.

It is very important to learn to efficiently use timing budgets of the system. Management activity plays an important role in the optimization of using time resources in management activities. It is known that only 1-2 per cent of the staff's general time is allotted to the performance of management tasks of strategic character, and it is considered to be insufficient [4]. Another point concerning the optimization of using timing budgets in developing the system is connected with the trend of "ups and downs" of the system potential. The pickup tendency means that the system is encouraged to perform its main tasks (aims). By this moment the system must be ready to use all its resources for receiving the best result. The downturn tendency, vica versa, opposes the system in fulfilling its main functions and, therefore, the system must perform miscellaneous functions for preparing pickup conditions.

The total potential of the system can be raised by its interaction with other systems. The condition of the system potential first of all is determined by basic resources.

As early as at the beginning of the XX century the prominent Russian scientist A.A.Bogdanov showed that the potential of the whole system considerably depends on the type of interaction of its elements (subsystems).

1.5.2. The Potential of "Well" and "Badly" Organized Systems

No doubt that technical, economical, political, military, scientific, intellectual, educational and other potentials of a complex system considerably depend on purposeful, consistent and rational relations between elements and on the rational organization of the system itself and its structure (economical, political, military, educational, scientific, etc.). It follows from this objective regularity that if the interaction of structural elements ($a_1, a_2, ..., a_n$) of system A are purposeful and consistent or synchronized, the system should be considered to be "well" organized. The more purposeful and consistent the actions of the system elements are, the better organized the system is. In a well organized system under emergency the potential P of the system A manifold exceeds the sum of potentials of all the constituent elements (subsystems): $P(A) > [P(a_1) + P(a_2) + ... + P(a_n)]$.

As for the entropy of system A, it is less than the sum of entropy of its elements because of accurate and consistent interaction between the system elements. If in the process of integration the entropy of the system decreases, it means that a new integrative property of the system emerges which did not exist before the elements were integrated.

The above-mentioned example corresponds to an overadditive non-linear addition – "the whole is bigger than the sum of its parts". The interaction of cells and separate organs (elements) in a holistic healthy organism (system) represents a simple example of reasonable and interconsistent interaction.

Under emergency, (A) acquires qualitatively new integrative properties, which is indicative of the overadditivity of the system.

Thus, if the degree of organization of the system, for instance, of the country as a system is high and the interaction of its subjects is interconsistent and purposeful, then the potential (economic, scientific, military, etc.) of the country is much greater than the sum of potentials of its constituent subjects, and the entropy is smaller than the sum of entropies of its subjects. Similarly, in a well organized organism, chair, laboratory, department, institute during the discussion of new problems there emerges new knowledge, a new solution to problems which separate members of the team did not have, therefore, the team's knowledge (intellectual potential) is greater than the sum of its separate members' knowledge. Knowledge gives birth to new knowledge, and co-ordination of interactions of the group's members multifold improves the effectiveness of each of them. Similarly, a duet's, a trio's or quartette's performance is much more impressive than a solo performance of each of these brilliant musicians.

It follows from A.A.Bogdanov multiplicativity law [5].

In neutral or pseudosystems, where the degree of organization does not ensure effective and co-ordinated interaction of elements, the potential of the system is equal to the sum of potentials of its constituent elements

$$P(A) = [P(a_1) + P(a_2) + ... + P(a_n)].$$

As for the entropy of system A, it is equal to the sum of entropies of its constituent elements. Consequently, superposition or emergency is characteristic of such systems. Such simple summing up (addition) takes place in such parameters of the system as, for instance, energy, mass, etc.

In badly organized or disorganized systems or rather pseudosystems, when the interaction of elements is uncontrollable or random (chaotic[1]), the potential of the whole system is equal to the potential of its separate averaged element $P(A) = [P(a_1) + P(a_2) + ... + P(a_n)]/n$ or

$$P(A) = [P(a_1), P(a_2), ..., P(a_n)].$$

In a badly organized system or pseudosystem, when the interaction of its elements is antagonistic, and each element of the system opposes all the rest of them ("war of everybody with everyone"), then the potential of the system is less than the potential of the weakest element of the system, and the entropy of the system, vice versa, is greater than the entropy of the smallest element of the system

$$P(A) < \min[P(a_1), P(a_2), ..., P(a_n)]$$
$$E(A) > \min[E(a_1), E(a_2), ..., E(a_n)]$$

In this case the potential of the whole system is estimated as the difference of potentials of elements in a random coupled order until one element of the system is left. And the less differentiated the values of element potentials are, the lower is their resultant system potential $P(A)$ and for all the known distributions in extreme case when $\tilde{n} \to \infty$, $P(A) \to 0$.

From the foregoing it follows that there exist "badly" organized systems or pseudosystems which in strictly scientific sense do not meet the system requirements and first of all its integrative properties. The potential of such "bad" systems or pseudosystems is less than the sum of potentials of their constituent elements and even of their one element.

From the foregoing it follows that if, for example, an independent state as a system has badly organized economy, science, education, etc., the interaction of the subjects is badly controlled or uncontrolled and antagonistic, and the potential of such a state is low and may be even lower than one subject's potential.

[1] chaotic means that the behavior of any element in the system does not depend on the behavior of all the rest elements and separately of each of them

1.6. Background General-System Regularity or "Background Principle"

The "background regularity" or "background principle" makes it possible by means of changing background radiation to reveal the object. It is of universal character and it is widely used in engineering, biology, medicine, meteorology, socio-economic environment, linguistics, etc. Background is a system's attribute. Backgrounds of systems are often presented by stable processes ensuring the functioning of the law of composition of system relations. The background regularity of revealing the prototype system consists in the fact that the background radiation is functionally connected with the object and, therefore, it is possible to judge about the object by studying only the background signals or its condition.

The background regularity consists in functional dependence of the background signal change or background condition change on the object's influence. When the background principle or background regularity is applied in the most general case, the system is presented as consisting of an object, a background, an observer and their relations. Let us take a simplified example of the object's influence on the background. A usual family consisting of parents and children may serve as a background. The members of the family as a result of adapting to each other over a long period of time form the family's habitude or a background radiation characterized by its own radiation. If an alien, for instance, as a son-in-law or a daughter-in-law, enters this family, it will introduce a certain disturbance into the established family, and by the changes in the condition (radiation) of the family (background) the observer can judge about the object. Thus, instead of studying the object itself, when it is difficult, it is possible to study the changes in the background condition (radiation) of the object, and by these changes to judge about the object. The "background effect", as a rule, results from the interaction of radiation of moving and stationary objects in conditions of coherent reception of this radiation. If the observed object is moving against the background of the radiating stationary object (background), then in the conditions of coherent reception of radiation a background effect manifests itself in oscillatory processes or in sharp temporal changes in the background radiation. These oscillatory processes or temporal changes depend on the geometrical and dynamic parameters of the moving object, and do not depend on its scattering power, i.e. visibility. Thus, objects, "invisible" for traditional observation, can be revealed and their parameters can be defined if there is a possibility of coherent registration of background radiation.

The background regularity is thoroughly studied for the purpose of revealing "invisible for radars" moving objects such as aircrafts, tanks, ships, produced according to the "stealth" technology. The background regularity of revealing moving objects was established by I.V.Pranguishvili, A.N.Anuashvili and V.V.Maklakov and in 1983 and in 1996 a diploma was issued for opening a forum which runs as follows: "The unknown before regularity of display of object mobility based on the radiation of a stationary background has been established: the intensity of the complex radiation amplitude averaged during the observation period at the observation point is reduced subject to the size of the travel of the moving object during the observation time, and if the travel of the object happens in the range of the radiation wave length, then the mentioned intensity is reduced from the radiation intensity level of the moving object to an infinitely small level and does not undergo a considerable change during the travel of the object in the range of its size, and then works for the radiation intensity level of the stationary object".

It is traditionally considered that a useful signal can be received from the object itself, and that the background radiation represents noise, disturbance, and it must be suppressed, otherwise the background disturbance masks (camouflages) useful signals coming from the object. According to the background regularity, the background radiation is functionally connected with the object and, therefore, we can judge about the object by studying the functional dependence of the background (noise) signal on the object signal. It is especially important, when the background radiation (traditionally noise, disturbance) is much greater than the traditional useful signal.

The background principle allows to improve the existent radio-, light-, and hydrolocators and to increase their efficiency in revealing unobtrusive moving goals including those produced by means of the "stealth" technology. This principle can be used for protection goals, for the security of airports, sea ports and other important objects. Besides, it may be used in scientific research in physics, biology, medicine, etc. [22].

The essence of the "background principle" and background regularity is thoroughly analyzed in [22] by example of revealing invisible objects in performing technical tasks, and then the possibility of applying them in medicine, biology and other spheres is discussed.

1.7. The Laws and Regularities of the Socialist Revolution and Different Models or Stages of Socialism

A disregard of the objective laws and regularities of natural, social, economic and other systems in decision-making and practical actions results in negative consequences sometimes because of the lack of objective prerequisites for the planned transformation or the lack of its correspondence to the objective natural and social laws and regularities or even its being in contradiction to them.

Did the socialist revolution of 1917 take place according to the objective social laws and regularities or did it happen quite contrary to them as there was some deviation from the objective laws and regularities of transforming a social order? V.I.Lenin in his polemics with Sukhanov admitted that in 1917 in Russia there were not sufficient but minimal economic and cultural prerequisites for socialism. Therefore, in absence of the required regularity or sufficient preconditions for a socialist transformation and, consequently, for a socialist revolution we received a voluntaristic model for developing a socialist country. This model was a deviation from objective regularities and, consequently, some authoritarian correction of the objective course of the world history. Therefore, the planned revolutionary social reforms did not correspond to objective laws and regularities and had to be compensated for by a powerful organizational factor.

It should be mentioned that such notions as justice and injustice are more closely connected with ethics, moral code of the society and not with objective regularity.

If we divert from some "prematurity" (untimeliness) of the socialist revolution in Russia as in an undeveloped country, where there was no economic and cultural basis for developing a socialist social order as a fairer one than capitalism, then we may state that the first decade after the revolution, the first phase of socialism was developing according to the "objective law". It spared Russia from catastrophe with which at that time the society was threatened.

By 1917 in Russia a great deal of the people's perturbation energy had accumulated because of the exorbitant oppression of the labor by capitalists and agricultural landlords, heavy casualties, and despondency. It was the people's perturbation energy which became the original cause of the socialist revolution in spite of the fact that Russia as a poor country with semi-literate population was not ready either economically or culturally for the socialist regime as a more progressive and fairer one. This social order required a stronger economy and a more developed culture.

As a result of the socialist revolution, at the first phase of socialist transformation the first model of soviet socialism, the so-called "military communism", or authoritarian-utopian socialism was established. Later, under the pressure of general devastation and mass discontent with "military communism", the new economic policy (NEP) emerged as the next stage of development. With NEP came a transitional model of socialist society called authoritarian-heterogeneous socialism [10].

The essence of the authoritarian-heterogeneous socialism consists in preferable development of the socialist sector and temporary limitation (or strict control) of the private-capitalist industry. As V.I.Lenin put it, NEP and, consequently, authoritarian-heterogeneous socialism, which to some degree eased the people's economic life, was introduced for a short time, however, in practice, after V.I.Lenin's death fear was aroused that NEP might cause the degeneration of socialism.

NEP was artificially cancelled and, therefore, a partial reversion to the first phase of the socialist revolution took place, and a new social order, the so-called authoritarian-liberalizational immature socialism with totalitarian distortion emerged. The cult of Stalin was formed and accepted by the people and served as an effective means of the authoritarian undemocratic rule.

Such socialism of 1940-50-ies was immature, primitive, undemocratic, and inhuman. The social ownership became state-bureaucratic; private ownership was cancelled, but necessaries of life, plots of land attached to houses, etc. remained in private ownership.

The authoritarian model of socialism played a positive role, it ensured the victory in the second world war, helped the country out of the postwar devastation, restored the ruined national economy, did away with the USA nuclear monopoly, ensured security of the country and the socialist camp. Therefore, the authoritarian-mobilizational model of socialism fulfilled its task and exhausted all its possibilities and lost its historical justification. However, by inertia it continued its existence. In the 1960-70-ies the country needed radical changes, general democratization in all the spheres of social life, there was an urgent need to raise the living standard, to ensure human rights and freedom of speech, etc. However, the dogmatically disposed political leadership obstinately refused to perceive it [10].

The unwise political leaders thought that it was possible to use the existent smoothly running authoritarian-mobilizational model of socialism first for entering developed socialism and then communism. In practice, the second half of the soviet socialism was a process of forming a second authoritarian-bureaucratic model of socialism. Its achievements did not correspond to those of other socialist countries. Under this latter model there were a lot of drawbacks. The main scientific and technical achievements were not in demand, therefore, the GDP and economy growth rate in whole were low; the people's abilities and initiative were not properly used; the equalization tendency prevailed and there was no differentiation of incomes subject to working data; ordinary people, in fact, did not participate in the state, region, district, and even factory administration; ordinary people did not trust the authorities any longer.

Thus, the authoritarian-bureaucratic model of socialism which was formed in the 1970-80-ies was irregular, abnormal and historically random, which shook the people's faith in socialist models and their perspective. The people relinquished their belief that it was possible to ensure worthy life for people on the basis of social ownership.

In truth, a "historical deadlock" is not socialism, but its bureaucratic distortion (model) which for 30 years delayed the emergence of a regular model of new non-bureaucratic, planned-market, democratic, people's socialism. In 1980-1988 such a model of people's socialism was in the air and in some professionals' minds. However, during "perestroika" (restructuring) the leadership of the country evaded the formation of this model of people's socialism.

During the first two years of the six years of "perestroika" (1985-1991) almost everybody hoped that real "perestroika" of the country and society would take place. People really believed the slogans: "More socialism – more democracy", "Socialism with a human face", "Less red tape in administration" and, having longed for changes for a long time, welcomed all this with enthusiasm. At first people believed that the party had elaborated a concept of "perestroika", though they had not seen it in mass media. During the following two years (1987-1989) the belief and hopes connected with the perestroika gradually weakened; in 1989-1991 doubts concerning the promising perestroika were becoming deeper and deeper, and the disillusionment resulted in general indignation of the majority of the people, because they realized that they had been deceived and sold out. Only those people were triumphant who thanks to perestroika grew rich without any difficulty. They promoted diametrically opposite slogans such as: "Democracy without socialism", "Dictatorship for capitalism", "Capital must hire administration it needs", etc.

It is obvious that in the history of perestroika from its very beginning the activities aimed at transformation of social order and restoration of capitalism were circumspect and systematically pursued by the political leadership assisted by domestic antisocialist and foreign (first of all, USA) forces. The restoration of capitalism in Russia for the sake of masking is called the transition to a market economy or return to the world civilization.

In the process of accumulating original capital in Russia the class of bourgeoisie was formed from plunderers, profiteers, bribetakers, people who free of charge privatized enormous state property. The first shadow soviet business was in fact legalized by passing the "law on cooperation" in 1988 and conditions were created for laundering criminal capitals. In the1990-ies hundreds of commercial banks were established, and criminal capital poured there.

Simultaneously, on a large scale the process of antisocialist ideological propaganda started in society. A great number of mass media acquired an antisocialist orientation which was in every possible way promoted by the then "democrats". They began to fundamentally deny socialism, inviting people into capitalism or into "modern civilization". The people were being convinced that socialism had finally proved its bankruptcy, and Swedish or Scandinavian socialism was presented as neocapitalism. A lot of people, especially young people, disappointed in the material welfare of socialism, were ready for the sake of good earnings to receive any local or foreign capitalists.

In 1990 the process of artificial disintegration and destruction of economy, decline in output, destruction of the foundations of social life started in the country. The breakup of the USSR (1991) caused a general systemic crisis, and the people faced the problem of physical survival. Private capitalist ownership was legalized. At the same time the thesis of "equality of all forms of ownership including private ownership" was promoted, and in the middle of 1991 a law on denationalization and privatization was passed. This law actually gave the green light not to equality but to the predominance of private ownership [10].

If V.I.Lenin's NEP implied that the socialist sector would retain its leadership, and private capitalist entrepreneurship would be temporarily allowed to a limited extent and strictly controlled by the socialist state, but Gorgachov's NEP on the opposite stipulated that the private capital would have dominion over economy and afterwards assume political power.

After the August of 1991 and the breakup of the USSR radical reactionary forces seized power under the leadership of B.N.Eltsin and aimed at destructing the socialist economic basis and replacing it with a capitalist one which caused degradation of the national economy, decrease of the living standard of the majority of the population by two thirds and of manufacturing output by more than 50 per cent.

A question arises concerning the main reasons for the USSR catastrophe and its scientific-historic explanation. It is obvious that the soviet society had not reached the stage of regular irreversible long-term evolutionary development, i.e. the socialist revolution had not had time to go over to the third phase or stage because the authoritarian-bureaucratic socialism was not that necessary stage. The fundamental reason for the USSR catastrophe was the fact that the authoritarian-mobilizational socialism had not been transformed into the necessary model of planned-market democratic people's socialism.

A number of distortions of authoritarian-mobilizational or authoritarian-bureaucratic socialism made it possible to compromise the idea of socialism for a certain period of time and its enemies managed to deprive it of the people's support. If at the next stage of its development socialism had coped with the task of being transformed into the necessary model, history would have pardoned it these sins. History had been expecting this transformation for decades but instead of it received this almost useless apparatus-bureaucratic model of socialism, and then the deceitful perestroika and the breakup of the country followed by a sharp decrease of the people's living standard. As a result, history and objective regularity sentenced the soviet authoritarian-bureaucratic socialism to death. If at the next stage of its development the socialism had coped with the task of its transformation, the elite would have forgiven it its sins [10].

It is known that the society's economic and social-cultural activities are subject to corresponding objective laws and regularities. The policy of building socialism is the more successful, the more completely it is guided by these laws and regularities. Political struggle and armed struggle as its extreme form have their objective laws and regularities as well. The laws and regularities of socialism construction and those of political struggle may not coincide. Constraints imposed by history on the whole process of positive socialism building requires that during the forthcoming period preference be given to the objective laws and regularities of political struggle.

Thus, summing up, we can state that socialism in the USSR was a baby born as a result of preterm labor in an undeveloped capitalist country. Then it may be stated that NEP was a brilliant way out of economic and social dislocation, a good way for rehabilitation of the national economy, but was of no use for constant policy [8]. NEP was unable to speed up the development of socialism. Such useful achievements of NEP as commodity-money relations, self-financing went over to the new economic mechanism and are being adapted to their new tasks.

1.8. Objective Systemic Laws and Regularities of Infomedia (Information Environment)

In the information environment there operate a number of objective laws and regularities.

The first objective systemic law (regularity) is the law of organization and limitation of information in social systems. The higher the level of organization of the social system is, the higher the level of limitation must be.

Under existing conditions when the process of globalization of economy, knowledge, trade, etc. is taking place, it is very important not only to satisfy people's and society's information needs, but also to protect persons, society and state from information threat and to provide information security.

It is evident that the creation of interconnected information space should precede the creation of interconnected economic and legal space. The interconnected information space becomes the backbone feature of any state.

The Okinawa Charter, elaborated by the members of G8 in Okinawa (Japan) in 2000, calls to bridge the international information gap [13].

An important systemic regularity in the sphere of information for complex processes (systems) is that the development of affairs considerably outstrips the process of perceiving them. Such was the situation during the perestroika in Russia. The processes were developing so fast that people including analysts had no time to understand what was going on. This lack of time for the leadership of the country to properly perceive the process of perestroika no doubt interfered with the correct management of the process.

Another objective systemic regularity for social systems is the regularity of information advance (lead) which means that the solution to informational interaction problems must outstrip any regular step in other spheres of social activities which makes it possible to better coordinate reforms, create favorable conditions for operating commodity, service, capital and labor markets, to ensure equal rights for acquiring property, using it and disposing of it. An interconnected information space must outstrip the creation of interconnected economic and legal space in the same way as it must outstrip making decisions in different social spheres. Therefore, information delay here will cause delay in other spheres of social activities, which is observed today.

In the sphere of producing information there operates the objective regularity of incomplete use of information including making decisions in management which is determined by both paradox of excessive information and lack of this ability in subjects [11].

In the sphere of information dissemination there operates the objective regularity of information distortion in its motion which is connected with subjects' different abilities and readiness to perceive it [15].

In the sphere of information consumption the objective regularity of compulsory expropriation and socialization of information which is connected both with unwillingness of subjects of their own accord to disseminate information and with the need for socialization of information in order to guarantee citizens' information rights, to ensure the creation of informational society and the development of the country's interconnected information space [13].

Three leading independent regional centers with their own specialization were established in the world according to international division of labor in the sphere of producing informational products and high-tech components.

The first center specializes and leads in producing program products for wide use in new computers. To this center belong the USA and Canada.

The second center to which Western Europe belongs specializes in the sphere of communication. Thus, Finland is in the lead in the sphere of producing cellular radio communication.

The third center, to which Japan and the countries of south-eastern Asia and China belong, has cheap labor and production facility and supplies components for computers.

Russia for the time being plays the role of a consumer in this international division of labor. However, as Russia has many expert computer people and highly skilled specialists in the sphere of information technology and thus it has a chance in the nearest future to play a leading role in the world in the sphere of producing systems and applied programs and in the sphere of new information technologies and knowledge-intensive products.

The informational sector as a new promising sector of economy and management basics must become quite attractive for Russia and CIS, where they have highly-developed science, education and a great number of experts capable of competing on the world information market.

Production of competitive means and systems and of information, telecommunication and communication should be developed in Russia; it is necessary to create favorable conditions for Russia's participation in international cooperation of producers of such means and systems. It is necessary to support fundamental and applied research and design in the sphere of information, telecommunication and communication.

1.9. Globalization and Integration – Regular Objective Proccesses Comprising Both Positive and Negative Factors

Globalization is spread all over the world and represents an objective regular process which shows in a number of spheres, mostly in the integration in the financial-economic sphere, in trade, culture, mass communication and information technologies. In future it will apply to (spread on)? other spheres. Globalization process is developing due to the objective necessity of the world structure and not due to production and spontaneity.

In the process of globalization the interference between different world-wide problems and social-economic as well as socio-cultural problems and processes of a concrete community takes place.

The idea of globalization lies in the human nature and influences it on the subconscious level, showing in the man's striving for the search and elaboration of reference patterns in the sphere of mutual relations. People aspire to explain the existing order of things in the world as well as the organization of the socium through a single whole: thus, in religion it is explained through the God, in mythology – though the Nature, in science – through knowledge, in politics – through power.

It should be noted that globalization of contradictions (oppositions) is, on the one hand, a novel phenomenon, but, on the other hand, it is a destructive one. Its conflicting (contradictory) nature shows in the fact that global economy coexists with the atomism (scattered nature) and non-integrity of production, labor markets, political organizations and societies. Globalization has enhanced forced involuntary aspects of security of states and, at the same time, has created new threats.

On the one hand, globalization, no doubt, accelerates economic development of countries, makes for the achievement of lasting peace, development of democracy and even for the unification of countries (e.g. European Union); on the other hand, along with the globalization process, growth of unemployment is observed, rich countries become richer, and poor ones – even poorer, inflation level rises, different economic and social problems tend to appear; all of this leads to conflicts, to the growth of entropy and to total chaos.

It is certainly clear that the extensive development of trade relations between different countries, organizations, transnational corporations, development of ties and communications, spread of information means (space, television, radio etc.) leads to the closer contacts between countries in different regions of the world. However, globalization, according to some scientists, will lead to transnational economy, which will cause such changes that will destroy the existing economic and political relations before the time when new relations will come into existence and , in this way, overall confusion and chaos will follow.

Globalization process has its own peculiarities depending on the country where it takes place. In Russia, the influence of world tendencies is different in different spheres of human activity.

Globalization problems are of great importance both for the whole world and for Russia . This country as well as other countries is connected to the cyberspace through the Internet, television and radio. As a result of this, westernization of Russia is taking place.

Globalization in the educational sphere leads to the unification and standardization of syllabuses and training methods, international educational projects, educational institutions, distance learning, Internet-learning, etc.

Russia will be able to enter the global space only by means of economy, trade (WTO), finances as economic and trade relations are governing conditions for this and they begin to influence politics. To include Russia in the international division of labor, it is necessary to set up international and transnational industrial and manufacturing corporations. Ideally, globalization presupposes equality in obtaining information, as well as international labour division, improving social-economic structure, spreading new technologies, advancement towards democracy, modernization of social life and bringing lagging communities up to the level of leading ones.

However, at present, we may observe huge inequality as regards the wealth of countries, possibilities of using the capital and information, the growing gap between productive capital on the one hand and commercial and financial capital – on the other, inequality between national structures, restriction of autonomy and independence of national civic institutions.

Some scientists and specialists develop a kind of euphoria concerning the rapid approach of "the globalization era" and, consequently, the improvement of world economy and world structure. But we should remember that in the 60-ies, in the West and in our country, there were overoptimistic prognoses concerning the development of economy; this was due to the development and use of atomic energy, microelectronics, automatization and robotization. A number of foreign scientists as well as the scientists in our country thought the 80-90-ies to be “the Golden age”, the years of plenty that would come through the continuous growth of power and nature exploration. The 90-ies were imagined to be the decade of general prosperity. But, as the reality has shown, optimistic prognoses did not come true. F. Fukuama, a well-known scientist and futurologist predicted that in the 90-ies the whole world would become unified according to the USA model and the overall well-being will set in. But the real life and the course of events have shown that these prognoses were only illusions and no general welfare and prosperity ensued.

1.10. 20% and 80% Law

Vilfredo Pareto (1848 – 1923), an Italian economist, formulated a principle (law) according to which smaller parts exiting within a large group or a set are a lot more significant as compared to their relative weight in this group.

Practically all business companies, while analysing their businesses, make certain of Pareto’s principle (law): only 20% of their clients bring 80% of the turnover. No matter how fast the partnership network is growing this ratio stays the same. Therefore, being “hostages” of this principle (law) both distributors and vendors try to do everytinh they can to strengthen the partnership with those companies that fall to the share of 80% of the business. The results of this policy are obvious – only 20% of the clients have and will always have maximum of support.

As 20% of clients ensure 80% of income, the policy of their support is carried out. On the other hand, rhere always exist the very 80 % of the companies that ensure only 20% of income. It is a well-known fact that 20% of scientists create 80% of scientific production while 80% create 20%. However, they cannot be separated from each other because if these 20% were separated from the rest of the group, then for the proper functioning of the system these 20% will divide into 20% and 80% again. Consequently, grate support should be given to 20% of scientists and not 5% or 40%. It is known that in germany, Czech Republic and othet countries 20% of people drink 80% of beer. It has also been determined that in commerce (especially in e-commerce) 20% of clients purchase 80% of production, so there are special benefits and special attitude with respect to them. This law is valid not only in our society but in nature as well (ant-hills, swarms of bees, etc.) A more detailed description of this law is given in [22].

1.10.1. Regularity of a Great Demographic Unbalance between Rich Developed Countries and Poor Developing Ones and Availability of their Access to Information through Mass-media (Television, Radio, the Internet, Press)have Created Hatred, Radicalism, Separatism and Terrorism; all this will Lead to “World War IV”

Our civilization is in great danger as a great demographic unbalance is impending; this unbalance may cause great changes in the society - the socium.

It is a well-known fact that in developed and civilized countries in Europe and America as well as Japan, there is no increase in population. The number of children in a family does not usually exceed one or two: the civilized family, which has been a moral "stabilizer" for centuries, is slowly disappearing. It is clear that technologically developed countries need fewer hands due to the high level of mechanization and automation of labor.

On the contrary, in poor developing countries with a low technological level millions of babies are born; when they grow up they become unempoyed. Poor people in Africa and in Arab countries have 5-6 children in a family; later they increase the ranks of the unemployed.

Beginning with second half of the XX century, when people living in poor and developing countries started to have acess to information through mass-media (television, radio, the Internet, newspapers, telephone and transport communications, etc.), they saw with their own eyes the inequality as regards the standard of life, the injustice and imperfection of the way the world is organized. These people came to know how rich and powerful these countries are; consequently, hatred and despair ensued which gave rise to radicalism, separatism and terrorism as a means of revenge.

Until the second half of the XX century, while there were no mass media, these people did not realize the unfair stratification of the society and so there was neother hatred nor radicalism or separatism and people did not consider terrorism to be the means of vengeance to the unfair world.

Part of the hatred, hostility and radicalism was mostly taken up by the dynamic development of the USA, Europe, Japan. However, at present, this absorbing reservoir, according to Bestuzev-Lada, is almost exhausted; therefore we are at the first stage of World War IV, which, unlike other wars will mostly consist of terror.

Poor states with excess demography are usually characterized by a tendency towards totalitarianism and external aggression. The aggression of Iraq against Kuweit with the attempt to take into possession Kuweit's wealth or, in case of Kosovo, to seize the land, may serve as a typical example. It was precisely the demographic pressure that ended in aggression. In fact, Europe (and Russia – as its part) is presented with an ultimatum to let this demographic wave coming from China and other countries spread all over the world.

Traditional religions are not as dangerous as sects that give rise to religious fanatism. International mafia is also very dangerous as it controls drug and arms markets, "slave-trade", prostitution and, which is the most dangerous, great masses of unemployed people. Countries like Tajikistan, Cechnya as well as the countries of Transcaucasia and Northern Caucasia, where three-fourths of the population are unemployed and, consequently, migrate to Russia in search of jobs, are very dangerous for Russia as regards terror, criminal situation, drug trafficking, etc.

Radically minded groups of people can take possession of nuclear, chemical or bacterial weapons on a small scale. Thus, a Japanese sect Aum Sinrike had a "rehearsal" for this in Tokyo underground.

Terrorists can use computer hackers for bank or financial systems.to become degraded. They may destroy the functioning of computer networks including the Internet and , in this way, cause great damage to the state by deranging the functioning of airlines, railways and causing the flow of funds failure.

Terrorists become especially active when they scare people or exert pressure on authorities. To achieve this it is necessary to spread information about acts of terrorism among population. Obviously, if such information were not widely spread, terrorists would lose interest in the acts of terrorism.

1.10.2. Regularities of Caucasian Mentality – Being Devoted to One's Clan and not Perceiving People Belonging to another Clan

It has been known from history that Chechen people and other peoples of Northern Caucasia are characterized by a special type of mentality which manifests itself in the devotion only to people of their clan. This objective regularity is confirmed by the history of these peoples. As the viable model of the state system and social order of this people is based only on the existing mentality, the most natural organizational structure for Chechnya will be a parliamentary republic like the one in Dagestan.

However, if after the referendum on Constitution, a presidential republic is established in Chechnya, then, for the purpose of taking into account objective regularities of Chechen mentality, it will be necessary to set up a council of elders consisting of representatives of different clans to approve all the decisions taken by official authorities (President, parliament, government). Since Chechen people, proceeding from their mentality, believe in and are devoted to authoritative representatives of their clans, they will implicitly fulfil all the decisions made by official authorities after they have been approved by elders. Therefore, it is advisable, along with the official government, to create an inofficial structure such as a council of elders for the approval or disapproval of decisions takes by power structures. Only such an approach to making decisions and their fulfillment is the right one because, in this way, the mentality of people and its regularities are taken into account.

1.11. Regularity in Phenomena Reversibility

As it is known, there exists a general systemic regularity in phenomena reversibility which is expressed in the method of three fields and three currents [24]. If we obtain phenomenon III through a certain combination of phenomena I and II (I+II→III), then variants can be reversed – the combination of phenomena I and III will result in phenomenon II (I+III→II), while the combination of phenomena II and III will give us phenomenon I (II+III→I).

There exists such a regularity in nature when phenomenon I causes phenomenon II (I→II) and vice versa, phenomenon II causes phenomenon I (II→I). For instance, heating of a body (I) causes its luminescence or electromagnetic radiation (II), i.e. (I→II) and, on the contrary, electromagnetic radiation (e.g. laser beam) (II) directed at a body causes its heating (I), i.e. (II→I) Similarly , gravitational interaction (I) causes the fall of a body (accelerated motion) II (I→II) and, inversely, accelerated motion (II) causes inertia force (I) which is equivalent to gravitational interaction (I), i.e. (II→I).

There are a lot of facts in medicine that conform to the regularity of phenomena. Thus, there are a number of cases when people were cured of cancer through auto-suggestion. They constantly convinced themselves that they were healthy and, in this way, symptoms of cancer disappeared and they were cured.

The above-mentioned real facts of self-recovery can be scientifically explained only on the basis of regularities in phenomena reversibility. In classical cases during the process of recovery certain biochemical processes (I) take place in the human body; in the end, they cause psychical phenomena – a recovered person realizes that he is absolutely healthy (II), i.e. I→II; but there are cases when, on the contrary, by means of psychical auto-suggestion (II) corresponding biochemical processes occur in the body (I) and a person recovers (II→I) [17, 24]. Thus, the standard or usual process of treatment brings forth a psychological process of treatment brings forth a psychological process I→II and unusual treatment by auto-suggestion causes a biochemical process II→I.

There is a well-known case of neurosis in medical practice when a patient (a mathematician) tried to convince his doctor that he had a frog in his (patient's) head. Later on, the doctor gave him narcosis and convinced the patient that he had operated on him and taken the frog out of his head by showing him a killed frog. After this the patient calmed down and stopped complaining of a headache. The patient believed in the operation; this confirms the fact that the psychological impact caused physiological phenomena in a certain part of the brain making unpleasant sensations disappear. In this way, due to the psychological impact in the form of suggestion, positive psychological processes may arise and a person gets cured. Afterwards, when the patient was told about the lie the pain in his head recommenced.

There are also cases when patients believed that a certain expensive, widely advertised medicine could help them recover. Then the patient is given aspirin and, owing to auto-suggestion, complete recovery takes place. In a number of cases a person under pain of death unconsciously jumped over obstacles. If he had had time to realize how high the obstacle was he would never have been able to jump over it. Being gripped by fear and having no time to realize the phenomenon, a person is able to overcome such obstacles and difficulties that are impossible to overcome deliberately. According to some literary sources, there were cases when a mother, quite involuntarily, was able to lift a heavy car, when her child was run over by it. In a usual situation she would never have managed to lift in alone, without any help.

All these and similar examples are reflections of a systemic regularity in phenomena reversibility, as we observe a reverse case – a person, being under the influence of certain psychological facts, changes and acts in an unusual way. The factor of auto-suggestion or not being able to realize a phenomenon or a process concentrates all energetic and spiritual reserves of a person, which helps to overcome great obstacles or to recover from a serious illness [24].

There exist special groups and psychotropic devices by means of which it is possible to control and manipulate the human psyche. These psychotropic means help to realize psychic energy a person possesses.

Regularity in phenomena reversibility lets us believe that in the future psychiatrists using modern psychotropic means will, with the help of auto-suggestion and some other special impact, be able to cure such grave diseases as cancer, AIDS, etc.

1.12. Regularity in Energy and Information Exchange

The man must learn how to coexist with nature and community without violating laws and regularities in energy and information exchange.

It often happens that for the purpose of getting an integral systemic picture of processes and events that are going on, one and the same problem is considered in complex from different points of view, and the regularity according to which these processes and phenomena take place (even accidentally) is thoroughly studied.

It should be noted that randomness is a succession of uncovered happenings that are beyond our understanding. The probability of the appearance of variates depends on certain laws, e.g. normal Gaussian or power laws.

In the East it has long been known that our world is much more complicated that its conventional picture perceived by our five organs of sense.

We are used to living in three- and four- dimensional space where the current time is the fourth co-ordinate of space. Having the ability to perceive information and thought only through four-dimensional mode, we can hardly understand the reality of processes that take place around us. Beginning with five-dimensional spaces, the fifth dimension (astral plane), time (t), as a co-ordinate of a lower metric space loses its meaning of a directed temporal flow. This means that if we consider physical processes happening in a conventional four-dimensional space through the astral plane (five-dimensionality) as Vanga did it, we will see an information field of events (IFE) in which any man is born, lives and dies. The main thing is that mass, time and distance don't play any role in it. Therefore, (due to the energy and informational law that does not take into account time, distance and mass) it is possible for a healer to diagnose a disease and heal his patients at any distance. This happens because the law and the regularity of energy and information exchange let a healer form an indefinitely large number of his own phantoms (astral-mental doubles) at any distance as well as in any period of time, even instantly. It does not make any difference whether the healer is in the same room with the patient or whether the process is broadcast on TV, radio or telephone, whether it is broadcast live or is recorded. In any case, correction (healing) takes place through IFE.

It follows from the law or regularity of energy and information exchange that when you remember somebody or talk about somebody, this person immediately appears beside you on the astral plane, and you appear beside him. This may be a living person, a dead person or the substance of another world.

It is known that any kind of knowledge can be used either for a good cause or to the detriment including - through astral karate. Owing to the astral stroke, a prepared adept can provoke an infarction, an apoplectic stroke and other diseases.

It is possible to summon a patient's phantom at a distance of thousands of kilometers either to help to destroy him. When we criticize (scold) somebody who is absent, we get into him on the astral plane and the person feels it in some way.

We know that thoughts and thought-forms are material. To achieve the materialization of a man's thought-forms, it is necessary that he imagine and see how it happens while being in transcendental state; afterwards he may forget about it completely.

1.13. Regularity in the Cyclic Recurrence of Starting a Global Crisis in the History of Society through Climate and Tectonics

According to some scientists [15] global crises on our planet are of cyclic character and, in Prof. Vladimir Trifonov's view, both negative factors – their climate and tectonics (seismicity) act synchronously. Deterioration of climate happens side by side with the rise in seismic activity. Both depend on the change in solar activity as well as on the angular velocity of the Earth's rotation.

The scientists studying the water balance of the Caspian Sea claim that the sea level tends to become lower or higher because of the climate. While investigating the tectonics of the Caspian Sea, the scientists noticed the following regularity: if tectonic foci in the sea become active, the sea floor goes down and, consequently, the water level also goes down; simultaneously the Volga run-off decreases. If, on the contrary, tectonic foci become active on the land, and not in the sea, the sea "contracts", the water level goes up and, at the same time, the river run-off goes up.

In V. Trofimov's opinion, climate and tectonics play a far more significant role in the history of humankind that it has been considered up to now. If there had not been three-year famine in Russia during the rule of Boris Godunov, a great disturbance would not have occurred, Boris Godunov would not have been dethroned and the history of the country might have developed in a different way.

It so happened that scientists studying global crises in the history of humankind often took a great interest in social, economic and political factors and did not pay any attention to natural ones.

The mechanism of rapid collapse of the Great Roman Empire and the simultaneous break-up of the civilization of the slave-owning system were also conditioned by the climate and seismicity. Thus, middle East and North Africa used to be main granaries of the ancient home. They supplied the Empire with bread and a lot of other products [13]. In the IV–V centuries BC a sudden deterioration of climate started – it became colder and more arid. The level of the Dead Sea, that used to reach Jericho, went down by 50 meters. Harvests decreased, which undermined food reserve of the Empire. The situation was aggravated due to the series of earthquakes in the Aegean sea and in the Eastern Mediterranean region; the seaports, through which Ancient Rome used to get supplies, were destroyed. After the collapse of Rome the slave-owning system gave place to feudalism. So, the natural crisis, by collapsing the Roman Empire and the slave-owning, system created a powerful impulse for global progress. This, according to some historians, is one of the basic laws of social development.

There were several similar climato-tectonic crises in the history of humankind. They recurred in cycles with periodicity of 1200–1800 years [15].

In the middle of the 3rd millennium BC deterioration of climate and a sharp rise in seismic activity caused the collapse of Tripoli culture in the Black Sea region and of other powerful communities.

In Tripoli people engaged in farming and thus they were equal. However, the deterioration of climate made them leave their homes, start a semi-nomadic life and search for the land of plenty. Forced militarization of the society began and chiefs appeared.

The next global crisis that redrew the map of the world and led to the radical change of the civilization of that time occurred in the second half of the 2nd millennium BC. Hittite kingdom collapsed owing to the invasion of wild tribes. Climate and earthquakes were the main reasons for this. The climate became hotter and more arid; this brought about problems with foodstuffs and, as a result, the unrest among the population. Another disaster followed – a destructive earthquake destroyed large cities of Hittite kingdom. The destruction of the cities caused the migration of the population, and a fierce struggle for new lands; those, who were armed better, won in this struggle. People started to search for new types of weapons. It was at that time that people started to use iron instead of bronze [16].

The last global crisis took place in England in the XVII century AD. The fall of temperature and change in climatic conditions caused famine. Food shortage led to disturbance. In Russia the disturbance caused by famine resulted in accession of Pseudo-Dimitri on the throne, afterwards, it led to his death. In Europe famine provoked religious wars and stimulated mass emigration to America.

It should be noted that nowadays most scientists agree on the fact that anthropogenic pollution is the important cause of the temperature rise on the Earth. The pollution appears because of the man/s activity – atmospheric pollution caused by industrial enterprises, motor transport, emission of greenhouse gas, in particular, that of carbon dioxide. However, there are different opinions about this problem.

So, some powerful empires and states appeared on the world map only to disappear from it completely. They existed for many centuries but then collapsed in only a few decades; the change in climate and seismicity were the main reasons for this [15].

1.13.1. Vagueness or Increased Entropy Zone

A number of specialists including S. Shoigy, Minister of Emergency Situations, believe that global warming or a sharp rise in temperature and rise in the entropy of the Earth cause all the natural disasters, including forest fires and floods in Russia and in a lot of European countries.

At Madrid intergovernmental conference held in 1995 UNO announced that global warning was a scientific fact [27]. Scientists give different prognoses concerning the intensity of global warming on the planet. Some of them think that a minor warming ($1{,}5^0C$) will occur and it will not have any significant influence on the climate. Other scientists and research workers of the UN International Council on Climatic Changes believe that the temperature will rise by 5–6 degrees, which will cause a real disaster. Firstly, great age-old glaciers will start to thaw and the oceans will overflow their shores. Japan will be the first country to suffer. Water will wash away all the coastlines on the Earth. Such sharp rise in temperature will destroy forests and lead to the increase in the populations of different insects; new lethal viruses will appear. In addition, global warming will cause great losses of agricultural production and other disasters.

All the scientists agree on the fact, that the most important reason for the global temperature rise and, consequently, entropy rise on the Earth, is the anthropogenic pollution; that means that only people and their activity are to blame for all the present and future disasters because people cause atmospheric pollution by industrial enterprises, transport, emission of greenhouse gas, particularly, carbon dioxide. However, this is a controversial question and a number of independent experts believe that only 30% of global warming occur owing to the man's activity, while the remaining 70%, may be considered a vagueness zone, which has not been studied yet, and so anything can cause such a sudden rise in temperature [27].

1.13.2. Global Uncontrollable Catastrophes

The Flood has been considered to be the biggest uncontrollable natural catastrophe on the planet during the history of the humankind.

According to Prof. E. Muldashev, the Flood was caused by the displacement of the axis of the Earth and the North Pole revolution by 60%. This displacement corresponds to the displacement equal to 1/6 of the Earth's circumference or to $\frac{40000}{6} = 6666$ kms, where 40000 is the Earth's circumference [18]. It should be noted that this axis displacement was not accidental and unsystematic; this phenomenon occurred according to a previously outlined plan.

The North and South Poles and the equator were located in absolutely different places than now; their displacement caused greatest cataclysms which are reflected in numerous legends about The Flood and Noah's Ark.

It is yet unknown what caused the displacement of the Earth revolution axis and the North Pole by 60% and why The Flood occurred; so it is unknown whether such a catastrophe will occur again or not; we can only imagine how it may happen [18].

In Prof. Muldashev's opinion, a new global uncontrollable catastrophe may occur in the future; this may happen if the revolution axis of the Earth and the North Pole shift to some region in the USA. In this case, Russia will become a tropical country, the new equator will pass through Tumen, Ufa, Saratov, Donetsk, Athens and will cross Sahara. It will cross the Atlantic Ocean diagonally and then move to the Pacific Ocean, passing through Micronesia, in the direction of Japan [19]. According to Prof. Muldashev, in this case, the coastline of the warm Arctic Ocean – Yakutia, Kolyma, Chukotka, Kamchatka, Taimir will become the best places for people to live. The Ob, Yenisei and Lena basins will be covered with tropical vegetation and New Amazon will arise.

There will be no winter in Moscow, Perm, Murmansk, Sweden, Finland, Norway. The whole territory of the USA and Mexico will be covered with a thick layer of ice, like Arctic. Owing to the rise of water, lowlands will be blooded. This will happen with Holland, Great Britain, Belgium, North Germany, Poland, Bulgaria, parts of Italy, Ukraine and France [19]. Probably, at that time the considerable part of the population will be wiped out by tsunami, as people will live mostly on plains, located below sea level.

Prof. Muldashev claims that due to the future displacement of the North Pole to the region of North America, there will be a new polarization of sun rays at a different angle than there was in the past when the Flood occurred. Consequently, the sky will become green instead of blue [19]. The analysis of the materials made by Nostradamus and Blavatskaya shows that in the period of Atlantis (Lemuritsa), before the displacement of the axis and the Flood, the sky was red, and the vegetation was purple-red. Nobody knows what kind of vegetation there will be under the green sky after the next global catastrophe, and what people will be like [19].

In seems that in the past there occurred quite a number of global catastrophes. Lemurine civilization perished but it gave way to the civilization of atlantes. In Muldashev's opinion, Aryan civilization originated in the heart of Atlantis civilization, which had already survived one global catastrophe – the Flood [19].

Evidently, if people could have predicted the coming displacement of the Earth axis, they would have taken some measures to survive, especially in the Tibet and the Himalayas [19].

1.14. Systemic Regularities in Power Élite Change

The main cause of power élite (i.e. of people invested with powers) change, is the change in aspirations of a society at a given historic period of development; élite should meet the requirements of the epoch [26].

Every epoch or historic stage of development has its own requirements towards the power élite. Therefore, power élite should correspond to the demands of the society at a given historic period.

Thus, in Napoleon Buonaparte's time, the society demanded that the élite consist of marshals and generals that had distinguished themselves in battles. In the epoch of imperial Russia a man belonging to power élite had to be well educated. During the first years of Soviet power educated government officials could speak several languages [26]. During the Stalin period of industrialization another type of people was needed - tough, ruthless, industrious workaholics, to whom the interests of the state were much more important than their private interests.

During the past ten years, since 1991, the country has gone through the period of collapse, disorganization, disturbance and theft, and so the demands of the time made to the officials are completely different. There are people who got "to the top" by chance because they were close to the "court", to the "family". They are unscrupulous, thievish, cynical and mercantile. They are not ashamed to seize up state property dirt-cheap, to deceive people. Hardly anybody could imagine that Pyotr Stolypin would seize up a plot of land for himself during the land reform. Nowadays, officials from power structures are not ashamed to do that even if everybody learns about their "deeds" from mass media. They are not afraid that sometime people will call them to account for their wrongful acts [26]. True, count Ignatov in his memoirs also mentions the facts as to how minor officials enriched themselves in Siberia at that time. Hertsen wrote a lot about plundering among Russian officials. It is evident that sin was always inherent to the officials and élite of Russia (least of all in the Soviet period). Thus, although Nikolai I made great efforts to stop plundering, all his attempts were futile.

It is clear that we won't probably be able to establish the epoch of creation and renaissance [26]. All this depends on how radical and great the forthcoming changes will be. Thus, for instance, if an epoch of creation, renaissance and prospective development of the country comes, absolutely other people should be in power and they will appear, their will correspond not to the epoch of collapse but to the epoch of renaissance.

Change of power élite depends on what we are going to create, what path the country and her leaders will choose. If radical positive changes start to develop, a radical change of the élite will take place. Everything depends on the degree of those qualitative changes that the country will have to go through. One thing is clear – the renaissance of the country and qualitative changes are hardly possible with this power élite, as it has privatized the power and has started to do private business at the expense of the matters of state.

Only a new power élite can meet the demands of a new epoch.

It is evident that the power élite of Russia must undergo a qualitative change. Public agents at all levels must explain causes of their concrete deeds to the people. Power structures of all levels should be more effective and should work in the interests of the people and not in their own one. The structure of government bodies should be simplified, their number should be reduced. Citizens of the country should have a real possibility to exert influence upon every official of the state. Citizens should have the right to declare a vote of no-confidence for any official.

Representatives of power élite should annually report back to their people and bear the responsibility for the deterioration of the people's living standard. If the basic social and economic indexes as well as the living standard deteriorate, power élite should be relieved of their posts and let more progressive power élite fill them, the élite that can find more effective mechanisms for increasing the living standard and life quality index of people.

Power élite should learn to respect people, take care of them. Authorities should learn how to listen to people and and how to hear them.

1.15. Science Entropization Process And Systemic Regularity of its Development

It is known that the increase in the true or new scientific knowledge forms an ever decreasing value per spent unit of cost as we are facing the process of science entropization or devaluation owing to reiterative descriptive, calculation, engineering, erroneous and other works that don't contain any new scientific knowledge. Therefore, the ratio of expenses on getting truly scientific knowledge to the value of different kinds of expenses may serve as a measure of efficiency for the scientific activity and may indicate the way of the most profitable use of expenses.

Research, conducted by D. Price, an American scientist, has shown that in the USA expenses on science grow in proportion to the square number of scientists or to the fourth power of the number of leading scientists [2]. For example, if the number of leading scientists in the USA trebles, expenses on science will increase $3^4=81$ times; this will constitute a huge part of the gross national product of the USA. On grounds of these data, D. Price arrives at the conclusion that science will inevitably be suppressed under the pressure of the social environment due to the great increase in science expenses, in the amount of information in particular specialization spheres as well as due to the limited nature of cognitive potential among scientists, extension of training periods, etc. According to D. Price, the initial exponential advance of science should slow down gradually, approaching a certain limit where it will stop without rising to absurd values.

A question arises – how to distinguish new scientific knowledge from the reiterative one, which is derived from something that is already known, and how to measure and assess scientific knowledge of different scope and level. Having processed a lot of facts, D. Price discovered an exponential (geometric progression) law of the increase in the number of articles, books, magazines, abstracts, scientific institutions, invention discoveries and expenses on research. But in the process of exponential growth of knowledge, the increase in the true (new) knowledge makes up an ever decreasing value per unit spent as the process of entropization or devaluation of science takes place; this process occurs due to erroneous, reiterative (plagiarism), descriptive and other works not containing new knowledge. A lot of articles, books, inventions do not give any increase in knowledge.

Science may be developing according to any law but entropic or devaluation processes hinder its development more and more. The main reason for this is the exponential increase in the number of research workers which leads to the decrease in their creative productivity. Science is getting more and more depersonalized, the number of collective scholarly works is increasing which makes it difficult to determine the share of each co-author. There exists a kind of lower limit to the scientific efficiency level below which a scientific institution starts to degrade rapidly [2].

Knowledge can be explicit and implicit (intuition, experience, skills). Explicit knowledge is based on logical thinking, analysis of events and is stated in books, articles, reports, etc.

Explicit knowledge can be divided into 5 categories:

1) truly scientific knowledge – this is structurized information on objective laws and regularities, cause and effect relationship between phenomena, information on facts that discover new phenomena and so on;
2) false or imaginary knowledge which does not correspond to the truth, e.g. mysticism and superstition and so on;
3) descriptive knowledge – information on facts that don't discover any new phenomena or regularities;
4) technical knowledge is presented in the form of calculations, constructions, methods and so on.
5) descriptive knowledge – information on facts that don't discover any new phenomena or regularities. Some people possess encyclopaedic knowledge but they cannot generate truly new knowledge.

It is known that scientific and technical activity of people is aimed at working out methods and means for the antientropic transformation of the nature and society.

In the first place, the mankind should try to cope with the "social entropy", which generates social inequality, antagonism and which disunites people, nations and states.

Rational life contains antientropic functions. The main condition for the existence and development of social systems, of animate and inanimate nature is to maintain stability which is achieved by two methods: 1. negoentropic or information method and 2. energy method. In the first case, the system reacts to external action in various ways opposing it through an appropriate qualitative reaction which is always new in every case. In the second case, the stability of the system is proportional to the energy of its internal ties.

1.16. Regularities of Non-linear Development of Social and Cultural Systems and Synergetics

Nowadays we do not have enough knowledge of laws and regularities concerning evolution, self-organization and self-government of complex systems i.e. the methodology of synergetic approach.

The problems of self-organization in social systems, including the culture system, are of paramount importance.

As synergetics deals with the study of phase transition in complex non-linear dissipative systems when the instability of structures increases or order parameters change, the synergetic approach secures the complex research of culture crisis and the dynamics of social and cultural processes.

Non-linear systemic thinking makes it possible to assess thinking not only by comparing its previous and subsequent states but by analyzing "volumetrical" spectre of system development processes. Scientists may compare the real course of subsequent events with the likely course of events at the alternative decision because synergetics makes it possible to realize the deep irreversible nature of developing its multivariance and alternativity [2].

The non-linearity principle is one of the basic ones in synergetics. The main idea of non-linearity consists in the possibility to have several (more than one) different solutions while solving non-linear equations. It follows that a great number of the ways of system development (evolution) correspond to the of equation solution sets.

When the influence of one system on another becomes too intensive, disturbance of structural integrity may follow. True, this may not ruin the system. However, if the transformations in separate structures become too significant, the system may turn into its new qualitative state having only lost part of its generic features.

Such transformations have already taken place in cultural system several times. However, the crisis situation did not lead to the complete destruction of the system; on the contrary, there appeared some kind of creative state for the more intensive development of culture [2].

Sometimes the process of change may be too fast and intensive according to quantitative indexes, so it seems that a new structure emerges through bifurcation of the developmental logic, which seems to be difficult to catch. However, this is not so. It is the non-linearity that imparts a particular meaning to social and cultural processes.

It has been determined, that the fundamental principle of non-linear systems behaviour is based on the periodic alternation of evolution and involution, unfolding and folding, implies explosion of activity and reduction of intensive processes to the stage of fading and weakening, centripetality integration and divergence, disintegration and even partial decay.

It is evident that the vector of the modern civilization is directed at integration or globalization and eventually all peoples or cultures will get involved in the global information – technological flow. It is hardly probable to avoid the pressure of unification and standardization of national traditions, maintain the isolated nature of ethnic cultures as their tribal sign (e.g. mountainous Chechnya or mountainous Dagestan) in the process of globalization. However, it is still possible to avoid typological uniformity. Signs of global culture are formed at the points of intersection of different cultures and later a kind of "mass" culture or self-development of another culture emerges, which is favoured by the globalization process. Nowadays it is impossible to exiust only at the expense of the cultural heritage of the past; it is also impossible to exist at the expense of the artificial forming of multiculture [2].

The history of culture abounds in examples on non-linear development. In some cases weak non-linear development of culture is observed, in other cases non-linearity is rather strong.

The process of development always presupposes the presence of periods of crisis and bifurcation. Stability of structures is secured by the balance between non-linearity and dissipation. A strong non-linear impact or too strong dissipation may destroy the structure.

It is known that every open system with strong non-linearity is characterized by pulsation, it manifests natural fluctuations and so the effort of the man to fully stabilize the interrelations in an open system such as "man-society-culture-nature" by means of powerful control is impossible. This would mean to deny the process of development, to stop evolution through the natural pulsation of the system and would lead to a crisis.

Not only interrelation processes between natural and social systems but also systems themselves are subject to pulsation or natural fluctuations. Therefore, being a complex open system, a cultural system does not need to be controlled either because this will impede the processes of self-development as well as the development of inner tendencies and systemic regularities. At bifurcation points a certain kind of predetermination is revealed [2]. True, some scientists believe that this way has already failed. They think that this may be a serious mistake that will lead to gradual decay (the repetition of Munich).

At present, social and cultural situation is, for the most part, characterized by strong non-linearity. It is known that cultural traditions of a society represent a peculiar framework of culture and the violation of the correspondence between traditions and innovations is indicative of its entry into the "aggravation mode" i.e. into the crisis phase [2].

World civilization as a united system undergoes uneven development in time and space and is constantly under the influence of integration and differentiation processes. A large-scale social and cultural unity is gradually formed; this occurs owing to the new integration phase connected with the globalization of economy, information field, globalization of interdependent economic ties. We know that the whole develops faster than its constituent parts. Therefore, it is more advantageous to develop together because this is connected with the saving of material and spiritual resources. However, even the most proper organization of structures into the united evolutioning structure leads to the moment of intensification and maximum development; owing to this, globalization cannot be total.

It should be noted that social and cultural dynamics is mostly accumulated in high-tech science intensive production, for knowledge and information turn into the basic productive resource. Therefore, the role of education and culture for the future reproduction of knowledge and information increases significantly. The importance of non-economic motives also rises. Moreover, on the one hand, the striving for technological process is observed all over the world; on the other hand, attention is directed towards the acceleration of production dematerialization process as well as to the increase in human resource importance. Non-linear development of the global world together with integration is armed at maintaining social and cultural differences [2].

In transition period a somewhat mild managerial influence and search for uniting the culture of values as well as radical changes in the way of thinking and people behavior are needed.

To overcome cultural crises spiritual reform and establishing a new system of values are necessary. A new system of values will, on the one hand, make it possible to unite the mankind in the face of common threats and problems and, on the other hand, to preserve cultural pluralism of the world community. Nowadays, the concept of "world culture" is aimed at creating such a world, in which various cultures, civilizations, religions and confessions could coexist in the atmosphere of mutual understanding, tolerance and solidarity [2].

It should be taken into account that during the evolution (development) of complex, powerful non-linear systems multivariance or alternativity of evolution is possible. In this case, it is impossible to predict a concrete way of system evolution out of several alternative routes at the bifurcation point, especially as it is predetermined by people. Similarly, it is also impossible to predict the way of the culture evolution as an extremely non-linear system at bifurcation (crisis) point. There are a lot of examples of cultural isocialism in the history of mankind when one culture suppressed another.

In practice we, as a rule, deal with very strong non-linear systems, therefore prognoses for systems development (e.g. weather, earthquakes, etc.) tend to become more and more short-term and can be realized only under certain conditions. Long-term prognoses for the development of systems with strong non-linearity are, as a rule, impossible.

In the present-day world self-preservation of different cultures that constitute the inner regularity of systems competes with the integration of world civilization conditioned by technological development and globalization of economy.

1.17. Problems of Ideality and Regularity of Non-Material Phenomena

Ideality is characterized by a certain magic "train" that makes it invisible but, at the same time, recognizable in spite of being invisible, ideality can be perceived by our organs of sense and can also be analyzed scientifically [12].

The ideality (non-materiality) phenomenon like phenomena of "spirit", "consciousness", "psychic", "information", "values", "ideas" etc, are covert, non-material essences; the visible material world reflects the invisible one.

Ideality (non-materiality) is still one of the mysteries of civilization. The concept comprised such substances as God, soul, spirit, consciousness, psyche, unconsciousness, sense, time, value, system, integrity, probability, information, regularity, quantum teleportation and so on. Modern science has not yet got common understanding of ideal phenomena but the fundamental nature that is the basis of non-material world, should be thoroughly investigated.

For centuries, the concept of ideality (non-materiality) has been present, either directly or indirectly, in different branches of knowledge in the form of such notions as ideality, harmonicity, synchronism, beauty etc. Essentially, there are fundamental problems in any science and they call for understanding the ideal nature of an object, phenomenon and processes.

Elaboration of the fundamental problem of ideality is essential for all spheres of knowledge as it affects principles of world order and world outlook. Ideality (non-materiality) is one of the most exciting mysteries of nature, the man, mankind, of the past, present and future of our civilization [12].

We surmise that ideality exists i.e. that there are certain non-material properties of matter or living beings in the material world. Thus, for instance, we know that the world is indivisible and integral, but at the same time, it is structurally diverse and so the diversity as the material property is one of the definitions of information and, therefore, information is ideality. The world and any other system are hierarchical and so, ideality is the property of the world's (system's) hierarchy and so on. In this way, ideality is the objective property of the matter which is actively manifested in the process of analyzing the properties of matter.

The problem of ideality in science concerns the elaboration of such theory of matter ideality that explains the most general regularities of the existence of non-material phenomena in the world [25]. In this sense, the regularity of the entropic equilibrium of systems, their decay and unification, their potential, their ladder-like and spiral development etc. constitute systems ideality as these regularities are very important objective properties of material systems determining their functioning and being actively revealed in the process of their study.

It is obvious that ideality does not represent the pure product of consciousness; in reality, it reflects the existing objective properties of nature and society.

Thus, matter is generated by ideal phenomena. Ideality is ubiquitous and it is the generalizing property of the matter.

For instance, in aesthetics the ideality phenomenon or theory coincides with the theory of harmony, beauty, golden section, morality i.e. the beautiful, harmonious, golden proportion, morality, customs are ideal.

In history, ideality is expressed in morality and purpose fullness of a historical process.

In chemistry, ideality manifests itself in such properties as periodic system of elements, catalytic reactions, etc.

In sociology, ideality is expressed in the theory of public opinion.

In physics, ideality is revealed in such properties as wave-corpuscle dualism, symmetry, conservation laws, etc.

In philosophy, ideality has been named in a lot of ways.

The theory of ideality is a system of hypotheses which is based on the duality of material properties, mirror symmetry of the world (light and darkness, left and right, plus and minus, a man and a woman, good and evil, particle and anti-particle, material and spiritual etc.).

Our world is integrally characterized by complementarity or unity of material and ideal sides.

Mutual reflection of material and ideal worlds is a very important question. Professor V.V. Nechaev has brought forward an idea according to which every material object has its own ideal image – an information icon and vice versa, every ideal object represented by the corresponding information icon may correspond to a natural or artificial material object. This mutual reflection (interreflection) or intertransformation of the material world into the ideal one or of the ideal world into the material one on the basis of modelling represents a fundamental problem of the universal nature of the substance ideality [22]. According to professor A.V. Panov, the system theory is the methodology of cognizing the ideal essence and the general system theory may be integrated into the ideality theory as its operational part.

The principle of material and ideal unity of the world or the idea of integrity or systemic character of thinking unites "western" and "eastern" philosophies [12, 22].

As research has shown, the concept of substance ideality may be used for the analysis of complex natural hierarchical systems as well as complex socio-economic systems. [22].

It is known that ideality as a phenomenon exists only in connection with human consciousness and that ideality is a specific "brain" form of reflecting materiality. Ideality, as well as materiality, is the fundamental property of matter.

The ideality factor is closely connected to the non-standard state of human consciousness. Thus, in [3] tying ideality to different states of consciousness is described and, at the same time, it is shown that ideality represents seven changed states of consciousness and six spontaneous states of consciousness.

In [3] it is shown that there exist only two types of the states of consciousness 1) consecutive state and 2) spontaneous state. Consecutive state, as the basic state of consciousness, in its turn, consists of: 1) basic state of consciousness and 2) seven changed states of consciousness.

Eight consecutive states of consciousness allow for only consecutive transitions of the previous state to the subsequent one without leaving out the previous stage; whereas the spontaneous state of consciousness appears spontaneously, allows for jumping over, is not controlled by us and can be interrupted owing to causes that are beyond our control. Eight consecutive states of consciousness are realized by the man when he is in his right mind and in psychically normal state. While the first or basic (normal, ordinary) state of consciousness is completely controlled by the man, the seven changed states of consciousness are controllable only to a certain extent. As regards the spontaneous state, it cannot be controlled.

These above-mentioned states of consciousness and their short descriptions are given below [12].

I. Consecutive state of consciousness
   A. Basic or ordinary, normal state of vigilance
   B. Seven consecutive changes of the state of consciousness
      1) Somnolence
      2) Sleep – can see dreams
      3) Autogenic state when the mechanism of psychosomatic self-regulation is switched on automatically.
      4) Objectification of subjective emotions, when a person "sees" his thoughts, feelings, mood, recollectione etc. as extraneous objects i.e. revealed moments of his inner life devoid of "my" quality (my thoughts, my feelings, etc).
      5) Changed state of consciousness, when mind is at peace and the switch of attention from a physical body to a "transparent" body occurs. This is like "leaving" a physical body.
      6) Changed state of consciousness – identified consciousness, when all realized conceived inner objects acquire "not me" quality.
      7) Changed state of consciousness when a mystic process of identifying yourself with light. (I), i.e. the observer is light. In Gospel according to St. John it is said "I am the light of the world", i.e. light is consciousness.

It should be noted that the basic or zero state of consciousness, autogenic state, somnolence, sleep switch in the indicated succession automatically. As regards such states as objectification of subjective emotions (state 4), "leaving" (emerging from) a physical body (state 5), non-identified state (state 6) and state 7 (when a mystic process takes place) these states can be reached by training the ability of a person to concentrate consciousness.

In state 7 a person reaches systemic equilibrium and stays in this state for only a moment, one, two or more minutes and then returns into the basic or zero state of consciousness [3].

II. Spontaneous states of consciousness are as follows:

1) Disconnection of consciousness.
2) Impaired state of consciousness.
3) Switched on intuition.
4) Prophetical consciousness.
5) Perception of other person's biofield.
6) Entry into collective memory.
7) Spiritual insight.
8) Shaman consciousness.

Transition of consciousness into different spontaneous states always occurs through moving from the standard basic or ordinary state of consciousness and this transition can be controlled by a person himself.

It should be noted again that all the diversity of objects and processes that are revealed to a person in spontaneous and changed states of consciousness belong to ideality.

When a person, through meditation, brings his psyche or consciousness into a special changed state, such state is sometimes called a "virtual" state of consciousness. Logical thinking is "switched on" and reserve possibilities of a person are stimulated. In the "virtual" state of consciousness the world as seen by a person, is bigger and more integral than in the usual state.

The man cognizes the "virtual world" through his imagination and starts to perceive the world created by his imagination.

When a person transforms (through meditation) into the object of perception, he views it not only with his eyes but by means of his other senses and he feels it from within.

The "virtual world" is a kind of a new parallel world. The technology of entering such a "virtual world" through meditation is well known. Therefore, any person can enter the imaginary virtual world through meditation, the world where there are no conflicts and where harmony and happiness reign.

It is very interesting to understand the way a person is scrutinizing some abstraction in painting, sculpture, music and how he apprehends it. It turns out that while looking at an abstract painting or sculpture, a person can see only those images or pictures which he are kept in his memory. Any abstraction is "sense chaos" i.e. the multitude of images (subimages), imposed on one another and pressed together. That is why every person takes in different images and pictures from the abstraction; this depends on the amount of information on the surrounding world kept in his memory. When a person brings his consciousness (psyche) into a special changed or "virtual" state of consciousness, logical thinking gets "switched off" and a person becomes extremely receptive to the images of *the outer and inner world.*

A prominent painter, sculptor, musician in the process of creative activity gets into a special "virtual" state and starts to apprehend a special world, connected with his *subconsciousness*. Subconsciousness is a kind of a "window" leading to some other imaginary, *"virtual"* world. He imprints "the imaginary" world on the *canvas* and this brings pleasure both to the author and to the public. Such kind of art is sometimes called "virtual" art as it is created in the "virtual" or changes state of consciousness.

"Virtual reality" is a new attitude and a new outlook. This is a new approach in thinking, vision (inner and outer) and feeling.

In *abstract* painting, sculpture that seem to be meaningless at first sight, people, with the help of consciousness and psyche perceive and comprehend and, consequently, realize only those objects and images on the abstract picture or sculpture that are in the right and *normally oriented state* and, therefore, can be perceived. Thus, for instance, we easily comprehend and perceive a person when he is in his natural and habitual state, particularly, when he is standing on his legs and not on his hands. Similarly, it is easier to comprehend and perceive a jug when it is in its natural state, when liquid is not being poured out of it. Therefore, if an abstract painting or sculpture is turned round its axis, a person can see much more sensible images *inside the abstract image* than while looking at them in their static position.

"Virtual" art which is based on the process of contemplation or creative illumination, gives great pleasure and the feeling of joy.

We should also say a few words about the correspondence between human consciousness and the *thought of a live hologram.* In 1970-s, Dr Carl Pribram, a well-known neurophysiologist from Stanford University (USA), suggested an idea, that thought is a live *hologram*; according to his concept, the mechanism of forming a "live" hologram in *the brain* is realized through signals having wave dynamics like recording wave field of light beams on a holographic plate ("Languages of the Brain", Moscow, 1975).

David Bom, an English scientist, in his book "Universe and a Hidden Paradox" (1980) showed that everything "that we see around us is nothing else than three- dimensional views, lying in the basis of objects". D. Bom supposed that the whole tangible reality, like a holographic picture, is *"a hologram" of an object.*

It is man's consciousness that generates all outer manifestations of a hologram, in the same way, as a hologram, while being broken into small pieces, reproduces all the images as on a whole holographic plate. D. Bom characterizes this deeper level of reality as an" inner order inherent in something hidden". It does not become apparent directly, it does not lie in the basis of each object constantly exposing itself at the generic level of objective reality and manifesting itself in laws of the physical world [12].

Rupert Sheldrake, a biologist from Cambridge, further developed the concept of "morphogenetic fields" ("New Science about Life"; 1982); according to this concept there exist *real force fields* which connect all the *cells* of a living organism. In his opinion, these fields are able to spread beyond the three-dimensional space and time and exist in a kind of "hyperspace". These fields secure the *connection* between the cells and the whole body and, virtually monitor the whole process of organic development. Sheldrake even stated that all organisms of the same kind are connected with morphogenetic fields [14].

Later on, biologists had to admit the existence of universal and *pervasive* ties in living organisms, so that the whole organism is "aware" of the state of each cell, and each cell "knows" the structure of the whole organism [14].

1.18. Regularity in Correspondence between 20 Archetypes of People and 20 Protein-Forming Amino Acids

The principle of *maximum entropy production at minimum expenses* is the primary principle of life evolution. In other words, the main tendency of evolution is the *stable increase in disturbance under the control of the environment* [14]. It turns out that this evolutional principle conditions properties of all chemical elements on earth. To ability of an element to cause *entropy* (for producing decay, chaos) is marked and explained by its position in the periodic system. For example, carbon has the strongest covalent bonds between its atoms as well as with the atoms of nitrogen, phosphorus, air, oxygen and others.

20 *amino acids* (protein-forming "atoms" of life) constitute unique natural complex complying with the principle of structural self-sufficiency.

These 20 amino acids are correlated with 20 human *archetypes* (stable dominants of the unconscious). In other words, there exists an experimentally established *bond* or correlation between 20 human archetypes and 20 different amino acids. It turned out that one of the amino-acids *corresponds* to each archetype [14]. Simply speaking, the predominance of a certain chemical element in an amino acid (protein-forming structure) determines psychophysical manifestation of human archetypes. So, the mankind is subdivided into qualitatively different archetypes of people. Twelve of them (by analogy with non-essential amino acids) are "intended" for the maintenance of metabolism on the levels of *morphogenesis*, *ontogenesis* and *arogenesis* [14].

An archetype determines the psychological model of person's behaviour. In archetypes a general plan of a personality type is reflected; this is A. Ukhtomky's dominant. The dominant reflects the evolutional level of thought-forming.

Eight "essential" archetypes out of 20 (by analogy with essential amino acids) are oriented to *allogenosis*, i.e. to the process of generating new qualities and ideas. Exactly these 8 archetypes, owing to their physical-chemical and psycho-functional peculiarities, are able to *take in* new information and generate it in the process of evolution (Shvanyeva I.N. Integral Psychology. Moscow, 2000).

For the representatives of these 8 independent archetypes the *ability* to take in and process information (information *homeostasis*) is a higher form of organization as compared with evolutionally necessary but already assimilated forms of homeostastis: elementary homeostasis (in ontogenesis), biological or microelementary homeostasis (in morphogenesis), energy homeostasis (in argogenesis). These independent archetypes are motive forces of evolution [14].

1.19. Systemic Regularities and Interrelations between Randomness and Predictability

Nowadays, it is vitally important to determine the connection between *regularity and randomness*.

The question whether all events taking place in nature or in universe are *predetermined* or some element of *randomness* is also present, has long been discussed by scientists [31].

In ancient *philosophy predetermination* was connected with order, with something that could be predicted, and *randomness* was considered to be a deviation from order or disturbance of order and thus, a negative phenomenon.

In reality *randomness* is *a chain of unrevealed regularities* that are hidden beyond the *threshold of our understanding.* The probability of random quantities appearance occurs according to the laws and regularities, for instance Gaussian (normal) law, power law or other laws.

A lot of scientists take an interest in the cause of randomness. In XX century, by creating the theory of *dynamic chaos* they came nearer to determining the causes of randomness. Indeed, the nature of randomness started to become clear only in the middle of the century when it was ascertained that *the instability of the steady motion is the cause or source of randomness.* Thus, for instance, if a spring flows through its usual course, the motion is steady and every thing can be predicted beforehand; but if the bed of the stream changes into a watershed, the situation becomes *unstable* as it is *impossible* to determine, where the stream will flow – to the right or to the left [31].

It appears that the instability phenomenon, as well as that of *randomness* is of fundamental importance.

In ancient times, *predetermination* and *cause-and-effect* relations were considered to be main categories, and *randomness* was thought to be a *deviation from order, its upsetting*, and was considered a *negative* phenomenon.

Students at schools and higher educational institutions are taught that the main task of science is to establish *cause-and-effect* relations by finding the cause of every phenomenon. True, for this purpose, the theory of *probability*, which is closely connected with randomness, is also used.

A discussion about regularity and randomness is still going on nowadays. A question arises - whether randomness "rules over" the world, or not; whether it can *save or destroy it* and whether *randomness itself is accidental or not.*

The type of bond between randomness and regularity is another issue of major importance. It has become especially urgent lately when the number of *natural and technogenic disasters* which are, as a rule, *identified with randomness*, has increased considerably. Destruction (ruin) of Atlantis, building the tower of Babel, wreck of the Titanic, the catastrophe in Chernobyl, the explosion on "Kursk", the act of terrorism in New York indicate that, in all probability there exists a certain *"rule or regularity for catastrophes"* and that their revealing is a question of primary importance to the science and civilization, particularly, in we take into consideration the fact that natural and technogenic catastrophes have become too frequent [31]. A lot of catastrophes are not exceptional, but at the same time, they are not *serial* but they are *recurring*:: some people, having learnt about a new catastrophe or accident, are sure that they are *exceptions to the rules.* However, there apparently exists a kind of objective *"rule of catastrophes*", which should be studied and taken into account.

To determine *predictability and randomness in science, nature and society, it is necessary to study both stable and unstable dynamic processes* which are united into random, chaotic and regular processes and form a general picture of the world [31].

Stable and unstable dynamic processes are axiomatically bound through Liapunov numbers and go over into each other. Therefore, we get a harmonious, *intregrated* picture of the world, uniting both random and predetermined or regular processes. Thus, while tossing a coin we cannot guess beforehand what we'll have, heads or tails, because we cannot foresee all the peculiarities of the initial conditions of tossing a coin.

1.20. Systemic Interrelations between Science and Religion, Knowledge and Faith and their Paradigms

The aim of science consists in *cognizing the truth* or, in any case, the materialist science, which recognizes the existence of truth. It is necessary to understand how the existence of God and religious faith, faith in God can help to cognize the truth.

The central point of modern religious doctrine consists, probably, in the fact that the Church *recognizes the role of reason (science) in cognizing the truth* but it considers that it is *impossible* to reach complete understanding of the *truth about the man and reality* (visible or invisible) that surrounds him only *by means of reason*. In addition to reason *revelation* is also necessary because the truth perceived through philosophical reflection and the truth of revelation cannot be mixed up because neither of them makes the other one unnecessary. In this way the Church recognizes science or reason to have equal rights with faith. Atheism does not see any place for God and does not know any proofs that faith in God helps to cognize the truth.

The Church considers that God himself put the desire to cognize the truth into the man's heart. Probably, the science should frankly admit that it has actually agreed to the fact that *ideal reality in nature really exists and that it cannot be reduced to only material reality.*

While traditional science uses *"reflexion"* as a method of cognition in such cases, non-traditional technologies use "meditation".

The contrast between *material and ideal, science and religion* limits our knowledge about reality. This is, first of all, formation of two types of thinking – *logical and "virtual*" (integral, relative). Therefore, to get an integrate picture of the world it becomes necessary to bring these two types of thinking together, to make them closes. Therefore, it has been observed recently that the "pure meaning" (science) and the virtual image of the world during the changed state of consciousness have become closer to each other.

Virtual state of consciousness sets in when a person brings his psyche (consciousness) into a special state, when logical thinking "switches off" and an imaginary, virtual world emerges.

Besides, it should be mentioned that nowadays, a great thirst for *faith and religion* is observed because there everything is more understandable for a person than the more and more complicated processes, created by the man in complex, artificial objects, as well as in socio-economic systems in towns, regions and countries because the complexity of an object or a process *grows much faster than the ability to comprehend, or understand them*; that's why the man passes to faith which is much easier to comprehend. Therefore, craving for faith and religion is also conditioned by the fact that people fail to understand and realize complex objective processes that take place in the material world.

Before we start discussing the role of religion and faith in God we should understand that *religiousness* is not merely faith in God, it means that a person belongs to some religion (confession) such as Christianity, Islam, Judaism, Buddhism and so on. A religious person believes not only in God but also in such key moments as the life hereafter, the immaculate

Conception, raising from the dead, Holy Bible etc. At the same true, a lot of believers believe in the existence of God but do not believe in the miracles of raising from the dead, in the life hereafter, the immaculate Conception. These believers are not *religious* people, they do not know the Bible, the Koran and other holy scriptures.

Such great scientis as Descartes, Newton, Einstein, Darwin, Pavlov, Cuvier and others believed in God. Thus, Newton was more proud of his theological works than of his works in physics.

Expert studes of 200 respondents' answers have shown that 60% of respondents are atheists and don't believe in the existence of God but only in the existence of nature with its laws and regulations studied by the science, 20% are *religious*, 10% believe in God but are *not religious*, 10% have no answer to this question.

A question arises – as to how a certain revival and upgrowth of religiousness in our country can be explained. There are several reasons for this; the main reason consists in the state policy with respect to religion and the spread of religious and, especially, orthodox ideology which shows in the support of Russian orthodox Church and in frequent use of the word "God," even in the national anthem of our country. Owing to the situation in our country, attempts are made to substitute the communist ideology by the orthodox ideology.

Why does it happen that in science some *paradigms* may replace others in case new, inexplicable phenomena and processes appear, while in *religion, unlike science, paradigms always stay the same (are invariable)*. Thus, it is known that in optics there existed the *corpuscular paradigm or the theory of light*. By means of this paradigm certain phenomena and processes could be explained. Later, it turned out that it was impossible to explain the phenomena of diffraction (dispersion of light) and optical interference with the help of this paradigm and so the *wave* paradigm of light appeared. Then it become clear that phenomena and processes in optic range could not be explained by these two paradigms and they had to be combined into a new *wave- corpuscle paradigm*. So in less than one hundred years the theory of light paradigm had to be changed three times to explain the results of new phenomena and experiments. Probably, it will be necessary to create still new paradigms in the future. Unlike science, paradigms in religion stay *unchanged*, they are eternal. Thus, for example, Christian religious paradigm which implies the faith in the existence of Almighty God, the immaculate Conception, raising from the dead, the life hereafter has not changed since the emergence of Christian religion (for over 2000 years ago) and may stay unchanged for many thousands of years in the future.

Thus, in contrast to religion there are no eternal paradigms in science and any science paradigm changes in time.

There is no common opinion among scientists about the interrelation between science and religion; and this question is widely discussed in mass media, especially in newspapers: A. Silin, doctor of technical sciences, thinks that science should recognize religion; on the contrary, Academician P. Gaidenko, believes that science and religion are incompatible. V. Ginzburg, academician of Russian Academy of Sciences, thinks that there are very few scientists that believe in God.

A. Silin writes: "Modern natural science does not need God. Nevertheless, scientists should admit that ideal reality in nature exists and it cannot be reduced to *material reality*".

At the end of the XX century the problem of the beginning of life on Earth as well as that of the man's origin became especially urgent. According to Charles Darwin's conception about the origin of more complex organisms, living beings as well as people appeared on

Earth by accident, due to natural selection, when *accidental changes* gave mutants a kind of advantage in struggle for existence.

Less adapted species become extinct and the winners, due to a qualitative leap, continued to exist.

It is apparent that the probability of accidental appearance of the man on Earth is negligibly small and practically impossible. There is not enough information, it just does not exist. So scientists, under the pressure of new facts, have *to turn to religion for help.* Cosmophysics consider that the appearance of man on Earth was planned by heavenly powers. According to A. Silin, science should admit that there exists ideal reality in nature and it cannot be reduced to material reality.

In acad. P. Gaidenko's opinion, science has approached such *limits* in issues of the origin of the world and the man himself which cannot be overstepped at the present level of knowledge; so a lot of scientists have to turn to religion in search of the answer to this issue.

Academician Alexander Spirin, a biologist, believes that scientists have no definite and clear idea about the origin of life and universe. The more we learn, the more questions appear and the usual paradigms fail to answer them. Academician Spirin asserts that it is impossible to make a complex device by means of evolution and constant improvement of a simpler one. Thus, for instance, it is impossible to make a TV-set by improving a loud-speaker. It is impossible to do without extra knowledge; in this sense, evolution laws are limited. A living creature, especially a man, cannot have originated as a result of evolution. The man is so perfect that it must have been created by a certain system, *"a deity"*, that is able to invent everything. We cannot find any other explanation for the fact that it when the Earth was too hot there was no life on it, but when our planet grew colder 4 milliard years ago, there appeared blue-and-green algae on it that had a perfect protein-synthesizing and hereditary apparatus. (A Spirin. A Step beyond Boredom. Obshchaya Gazeta №7, 14–20 February, 2002 г.).

A number of scientists, especially physisticts, think that our activity is limited by our knowledge and that any break-through beyond these limits makes us closer to the acknowledgement of superior power (the God). Academician B. Kadamtsev believed that it is not only the man's brain that is able to make a prognosis, to abstract or to have revelations. Universe is no less complex than a human organism so it can perform the same processes as the brain. Scientists from different spheres come to the conclusion (idea) of the existence of supreme divine powers because they have come across them more than once. This means that science needs the help of God and religion.

A lot of scientists try to combine science and religion in the sphere of control in physics, biology, chemistry and even mathematics. A number of regulations of the theory of relatively, cosmogony and quantum mechanics are so complicated and are at such variance with common sense that they border on mysticism and occultism. In P. Gaidenko's opinion, this has a harmful effect on both science and religion. Instead of conducting experiments science tries to find proofs of certain phenomena in miracle-working and does not need any theoretical prerequisites or a body of mathematics.

If science starts to aspire to understand religious dogmata and formulate rational constructions, then the very essence of the religious faith will be under threat. According to P. Gaidenko, for the sake of both science and religion it is dangerous to encourage the interpenetration of these zones of human culture. In this case, in P. Gaidenko's opinion, their self-destruction will be inevitable.

It is necessary to emphasize that these observations do not concern the faith of every single man. This is his personal choice, and nobody should interfere with it. It often happens that the spiritual part of a man's nature is not in contradiction with the intellect. So, a lot of great scientists believed and believe in God.

V. Kazutinski, doctor of philosophy, also thinks that science and religion are two different spheres of the spiritual activity of a man and from his atheistic standpoint, interpenetration of science and religion may lead to their self-destruction. In his opinion, Newton, Darwin, Pavlov and other naturalists, who believed in God, did not overstep the limits dividing *the rational and the transcendental.* God as a component of personal spiritual experience, stayed beyond the sphere of scientific knowledge. From Kazutinski's point of view, the origin of universe and the man does not need the transcendental factor to be involved. He hopes that the origin of the man will be explained by science. As the genofond of the man and anthropoid apes is similar up to 98% this shows that they are closely connected. However, the problem of the origin of human consciousness as well as its nature is still to be solved. According to this opinion, analytical psychology, which was created by Jung, is a science studying deep psychological emotions that belong to the sphere of religion. Human psyche, according to analytical psychology, includes both the level of rational thinking and those of the individual and collective unconscioues.

It should be noted, that strange as it may seem, representatives of many religious confessions prefer to abide by "God" only in its simplest and understandable sense. Actually, in the hierarchy (God-father, God-son, God-spirit or Holy Spirit) God-spirit is the highest divine spark which cannot be perceived by our intellect. And this something can be understood in any way, be it the Divine Reason or Logos [31].

First of all, it is necessary to acknowledge the fact of something existing above us, something that people should constantly be aware of. According to the theory of "Unusual Attractor", scientists working in the sphere of synergetics and applied mathematics have almost reached the fundamental bases of the universe and the Divine Reason or God-spirit [31].

It should be noted that the planet – the universe where the mankind lives, is an open system both for "good" and "bad" (aggressive) types of external influence; this may lead either to the positive organization and self-organization of the system (entropy decrease) or to its negative organization or degradation (entropy increase).

It has proved ratter unpleasant and even strange for politicians and authorities to acknowledge the openness of our world, especially in the sense of its Higher divine nature as this restricts their freedom of action in the face of some higher authority.

It should be mentioned that despite a considerable religious differentiation existing at this stage of development of the world community and different religioun, a certain synthesis of science and religion on the information level is observed.

Let us make a short survey of the formation and development of different world religions and determine how they are connected with the development of science [34].

Thus, since ancient times of polytheism, religion, through self-perfection and development of science, formed social world outlook and the corresponding picture of the world.

First, at the end of the 2-nd millennium BC Old Indian (Vedaic) pagan or polytheistic Veda or treatises appeared describing the cultural history of ancient India. Since that time paganism (polytheism, Gr. Theos-God) spread all over the world. However, eventually, representatives of religion, science and ordinary people started to realize that polytheism was a special kind of unity that formed integral world outlook or ideology on processes, natural phenomena and societies. The ideology of religious figures as well as scientific studies become the basis for such religions as Buddhism (Hinduism) and Judaism, Christianity and Islam that are connected with one another in the informational way.

Judaism originated in the 1st millennium BC; according to it, God is the only creator of the world and universe. At the first stage of its development Moses was considered to be the representative of God on earth.

In V-VI centuries BC Buddhism appeared in ancient India. Buddha (Gautamma) is considered to be the founder of this religion. Unlike Judaism, Christianity and Islam, in Buddhism God is not considered as the creator of Universe and the world; spirit and matter (physical body) are not contrasted with each other. Buddhism in India changed into Hinduism at the beginning of XII century.

A lot of peoples on the planet gradually shifted from paganism to Christianity, to the faith in God and his son Jesus Christ.

Islam emerged in Arabia in VII century AD. Mohammed, the prophet of Allah, is considered the founder of Islam.

In this way, the most widely spread religions in the world developed from paganism to one God, but to different faiths. It can be ascertained that at this stage of development all the religions have one thing in common - the faith in the Almiglty and Omnipresent God – the creator of Universe; this is the basis of each religion.

The main difference between present-day religions can be observed in the rituals of eulogies and celebrations in the name of God.

Religious figures, while studying phenomena and processes in nature and society unknowningly came to the conclusion that the Omnipresent God or, scientifically, information, is the basis of the world integrity. Thus, Pope Paul VI in his epistle to the leaders of Dominician order mentioned that Faith and Reason are in harmony with each other.

It is known that *Christianity* developed as one of the trends of Judaism at the beginning of our era. Christianity in its turn, became the basis for the development of other religions.

Nowadays, in spite of considerable religious differentiation, synthesis of science and religion on information basis is observed [34].

Different religions interpret the creation of the world in their own way. This concerns Judaism, Buddhism, Christianity and Islam as well.

More than three thousand years ago in Israel there appeared Tora, which was later transformed into the Bible consisting of the Old Testament and the New Testament.

In the first chapter of the Old Testament it is said, in the beginning God created the heavens and the earth. Now the earth was formless and empty, darkness was over the surface of the deep, and the Spirit of God was hovering over the waters. And God said, "Let there be light", and there was light. God saw that the light was good, and he separated the light from the darkness. God called the light "day", and the darkness he called "night"... and God said, "Let the water under the sky be gathered to one place, and let dry ground appear". And it was so. God called the dry ground "land", and the gathered waters he called "seas"... Then God said, "Let the land produce vegetation: seed-bearing plants and trees on the land that bear fruit with seed in it"... The land produced vegetation: plants bearing seed according to their kinds and trees bearing fruit with seed in it according to their kinds. And God said "Let there be lights in the expanse of the sky... to give light on the earth". God made two great lights – the greater light to govern the day and the lesser light to govern the night. He also made the stars; God set them in the expanse of the sky to give light on the earth…

God created the great creatures of the sea and every living and moving thing with which the water teems, according to their kinds, and every winged bird according to its kind…God made the wild animals according to their kinds, the livestock to their kinds, and all the creatures that move along the ground according to their kinds…Then God said, "Let us make man in our image, in our likeness"… God created man in his own image, in the image of God he created him; male and female he created them… By the seventh day God had finished the work he had been doing:: so on the seventh day he rested from all his work.

So, as we see, the solar system and life on Earth were created in six days. From the point of view of mythology, this is quite possible [34].

It is evident that science (knowledge) and faith (religion) are mutually complementary. Two different hemispheres of the brain are responsible for them. Thus, science can be considered a product of the left hemisphere, and faith, art, music, painting, intuition, images belong to the right hemisphere. Faith is a kind of love, and the scientists that believed in God, like A. Sakharov, thought that explaining the essence of the world cannot be reduced to matter and its laws.

Any human population or civilization that has no faith and religion, will inevitably die, as religion creates the moral and spiritual adaptability of population.

There has never been and can never be such people on earth who do not believe in something, be it a myth or a religious doctrine. Even in the times of the Soviet Union the myth about communism performed its function as faith in the bright future.

It is obvious that religion is one of the main forms of culture and according to the apt expression of scientists, science studies the "long" journey of the mankind while religion – its "mountainous" journey [9]. It should be noted that in the second half of the year 2001 a discussion was held on the article by prof. D.V. Pivovarov "Science and religion are commensurable (in newspaper "Nauka Urala"); in the process of discussion, most of the scientists advanced the opinion that faith is the highest form of spiritual perfection by far surpassing pure intellect and overcoming the "burden" of logical constructions.

In 2002 discussions on the interrelation between science and religion were held in newspaper "Vek" (prof. A. Silin, P Gaidukov, corresponding member of Russian Academy of Sciences, acad. V. Ginzburg took part in the discussions).

In 1992 a conference "Synthesis of Science and Religion" was held in California.

In 2003 a conference "Science and religion; knowledge and faith" will be held.

It is known, that religious systems came into existence together with the man and they are the integral part of his life. Religious emerge, develop and die, change their forms, but they are always present in the society. The reason for this lies in the very nature of these systems [9].

Lately, very interesting ideas about the connection between divine and earthly phenomena from the point of view of physicists have been published [9]. Exclusion of God from the structure of the world, nature and the man according to many scientists, may lead us up a blind alley. In the book "The Very Beginning" (Chicago, 1989) by a prominent physicist William Craig, it is stated that after analyzing all the existing hypotheses about the origin of universe we come to the conclusion that there exists a conscious reason that necessitates the emergence of universe; this reason is eternal and conscious, it created universe and time itself, at its own choice. The reason is eternal, the effect is not. In this way, universe could only have originated through the deliberate will of the creator, and then nothing will prevent us from calling him God [9].

Lately, more and more scientists start to think that universe, the Sun, the Earth, the man were created by God; however, we still don't know how it happened. It is still impossible to determine what place God occupies in our world.

An orthodox hieromonach Serafim (S. Rose) stated that people cannot see the location of God, Eden, hell and their inhabitants as they are beyond our spatial and temporal system, in another space (another dimension) that begins here but stretches in another direction.

S.Rose in his book "Soul after Death" writes, that the description of the journey reminds a journey from our three-dimensional world to the four-dimensional one, where the past, the present and the future exist as if simultaneously (like a time machine) and any body is imagined as extensive spatial structures from the moment of conception to death; in our three-dimensional world we observe the momentary temporal cut of the structures [9].

The characteristics of the four-dimensional world are so astonishing, that a lot of things that are impossible in our three-dimensional world are natural for the four-dimensional one.

There are instantaneous shifts at any distance and in space (as the speed of the shift is much higher than the velocity of light) as well as the free penetration into the closed space, "undoing any knots" and so on. That's why the presence of God in four-dimensional or five-dimensional space makes us understand the unlimited abilities of God to influence our three-dimensional world, people and their souls in a very unusual way. According to some scientists, God rules the world through substance which is sometimes conventionally called ether and which is an extremely thin substance that penetrates the whole world and is in weak interaction with substance but determines electromagnetic, gravitational, nuclear and other interactions [9].

It is considered that ether is that very bridge that connects four- and five-dimensional divine world with our three-dimensional one. Scientists believe that ether penetrates into our world from the fourth dimension where it exists in the form of ether gas, filling up the four-dimensional world like air fills the space above the Earth or radio-waves of the radar – the space under exploration. Waves of ether penetrate into our world perpendicularly to our space at every point, the great part of it returns to the four-dimensional space, taking away the information about the interaction with our world (by analogy with the information on interaction of the return signal of the radar). It is considered that ether is the perfect medium for passing information, at last in the form of electromagnetic oscillations; this is why it was invented by scientists of the last century (Faraday and other).

According to some scientists, ether is the substance that joins three- and four-dimensional worlds by means of energy and information channel. After interacting with the objects of our three-dimensional world and getting information about every point of the world, the ether flow at great speed gets back into the four-dimensional space where this information is taken to God [9]. This concept lets scientists explain the bond between body and soul, soul and God without resorting to such concepts as biofield and astral bodies which the man seems to have parallel with his soul [16].

It is assumed that religious man's soul is in the resonance with ether vibration thus getting spiritual food for his soul. Averting himself from God and religion, the man blocks commumication channels, loses food for his soul and thus grows weak, becomes brutalized. After death man's soul gets into the four-dimensional world, to God's repository [9].

1.20.1. Do People Have Regularity of Premonition Power

There have always been plenty of real, objective cases of many well-known people having premonition power; this power cannot yet be explained by science.

Steven King, a prominent American writer, while trying to explain on a logical level strange facts of some people having premonition ability, propounded a concept that our ancestors lived in natural, hostile environment that was full of dangers. One could not leave his cave without having some kind of premonition power. In the course of time, when life became more comfortable and less dangerous, premonition power became unnecessary, just as it was no longer necessary for people to have a lot of hair on bodies and arms. Therefore, this feeling only appears in extreme situations, but not always, otherwise people would not get into accidents, disasters connected with traveling by some kind of transport; they would simply refuse to go on that very trip or fly on that very plane.

To explain the fact that some people possess premonition power or the sixth sense, a lot of different experiments have been conducted and are conducted at present as well; however, they are neither strict nor very convincing. It is obvious that to examine these phenomena, which can shake foundations of many sciences and have a tremendous effect on fundamental principles of some branches of science, experiments should be strict and reiterative.

A lot of people, proceeding from their own experience, can say that their inner voice helped them to avoid danger; but science does not believe in premonition. The story about the hare that stopped Alexander Pushkin who was eager to get to Petersburg just before the Decembrist rebellion, does not convince scientists, they need a great number of facts and ever regularities, when the appearance of hares would save poets' lives in 70–90% of cases [16].

D.Staunton, an American psychologist, studied more than 50 cases of airplane crashes and 200 rail disasters to prove that some people have premonition power; he discovered some kind of regularity which consisted in the fact that on those catastrophic trips the transport was filled by 60% and not 75%, as usual. Some passengers were late, others were ill, somebody had returned his ticket, somebody had problems at the office etc; whatever the reasons, the facts showed that the transport was filled by only 60%. Analysis and research conducted by Steven King, a writer, showed that 16 passengers instead of usual 3 failed to turn up although all of them had bought tickets for the flight. In addition, another 15 passengers refused to take that flight whereas the usual number of cases constitutes half of this figure.

It is known that a lot of prominent people had premonition power. For instance, Winston Churchill, who used to sit beside his chauffer while traveling by car, once opened the front door of the car but then suddenly changed his mind and sat in the back of the car. Later on, while they were traveling, a bomb exploded in front of the car. Churchill said that while he was going to take his usual seat, he heard a kind of "groan" – "don't sit here", so he sat in the back and, in this way, stayed alive.

A film director Andrei Tarkovsky, while driving in the streets of Stockholm, suddenly asked the driver to stop the car, got out of it, came up to some spot and said, "This is the site of disaster". Several months later Ulof Palme, the Prime minister of Sweden was killed at just that spot by a terrorist [16]. A lot of people can remember how the "inner voice" helped them to avoid danger, but science does not believe in premonitions yet because there is no scientific systematization of facts and objective regularity has not been proved yet.

1.21. General Systemic Regularities in the Government Management System; Social Monitoring as Negative Feedback for Securing System Stability

State government systems like any other complex non-linear systems are sufficiently well described by the theory and laws of systems management.

By analyzing objective systemic laws and regularities as well as by generalizing social phenomena in a state or in a region, it is possible to work out a scientific basis of monitoring (control) and management of basic elements of a region or a state as a whole.

Let us consider in brief 4 systemic regularities that can be used for this purpose in the first place.

1. *The law of "unity and stuggle of opposites"* explains the reason for the ineffective use of national property and sharp decrease in social development. The meaning of the "unity of opposites" law is explained by the principle of "mirror reflection"; each opposite turns into the negative feedback which must provide for the stability and homeostasis of the system. In case of deterioration of the control (management) system of one of the interdependent antagonists (opposite), there appear elements of inter subordination instead of parity. Inadequacy between input output in the system of arguments disrupts the interdependence composition and, ultimately, the degradation energy that is generated due to the system breakdown, causes the secondary destruction. Weakness of bonds between the antagonists in the system may cause uncontrolled self-government of homeostats, similar to cancer cells; this has a negative influence on the state.

The social model of this regularity describes the mechanism that, according to the principle of transparency and openness, controls the interrelation both between the citizen and the society and between the society and the authorities thus imparting stability to the socio-economic system. At the same time, all the elements of the system (i.e. different state, public, economic structures as well as demographic, age and other social groups) and interrelation between these elements (e.g. competition, co-operation, mutual aid, neutrality and so on) should be controlled by the above-mentioned law of composition.

Social structures which represent the community of individuals whose behavior and interrelations depend on certain norms and laws, are the primary elements of the socio-economic system; on the other hand, state or ruling structures represent a small group of people that have been selected or appointed to govern the society and live at the expense of the society. Owing to this, the monitoring or control of these groups and reverse action from the society are the necessary conditions for securing the stability of the system "authorities – society".

If authority is the "mirror" of society, and society is the "mirror" of a citizen, then the negative feedback between them secures the stability of the whole system consisting of authorities and the society. This kind of feedback is best of all organized in a civic society.

2. *Critical point (moment) and crystallization center law* determines a critical point of transition to a new socio-economic formation and, correspondingly, to new mentality. This moment or point of phase change takes place when old interdependence or ties are weakened or destroyed and new ones have not formed yet or are not strong enough. At this very moment the maximum effect can be obtained by making minimum investments (crystallization center). Investments that are made before the critical point strengthen old ties and slow down

the process of transition to another state. If investments are received after the critical point, then even more investments and efforts are needed to ensure the transition of the system into the required state. Disregard of laws, corruption on the upper levels of power, a lot of errors, etc. have actually led to the crisis and postponed the "critical moment" when efforts of the community are still possible to secure the phase of transition to some new social system.

3. *"Systemogenetic"* law implies that any complex system (technical, biological, economic) must pass through all the stages of development; it is impossible skip any stage. True, it is possible to accelerate or slow down some stages, but they cannot be skipped otherwise the system will be defective or distorted; this is similar to the development of a fetus in the womb: during 9 months the fetus must pass through all the stages of its development, if this does not happen, a defective child will be born. It follows that for the proper development of capitalism in Russia it is necessary to pass through the first stage of primary accumulation of capital (the so-called gangster capitalism) however difficult if may be for the society. This stage evokes unfavourable criticism in respect to difficulties and injustice in the society. However, it is necessary to pass through this difficult stage of capitalism development in Russia. The situation is different in case of competent authorities and a lower level of corruption in the power structure; it could be possible to pass through this stage more quickly. However, in our situation, it seems that officials and the corrupted part of our society are interested in protracting the first stage of capitalist development in Russia and other countries of CIS. The "systemogenetic" law makes it possible to analyse all the evolutionary stages of the society development both in time of the Soviet Union and in our society, and discover irregularities at the stages of its development.

4. According to the empirical law of *"island effect"*, it is stated that powerful military, socio-economical, ecological and other systems cannot coexist with a much weaker and primitive system for a long time. In this respect, claims of, for instance Chechnya and Abkhasia on separatism and independence beside much more powerful systems means the violation of the objective systemic natural law of "island effect", which is impermissible. Therefore, if developed and more powerful states (like the USA, European countries) are interested, geo-politically and strategically, in a weaker state (e.g. Russia, Georgia), they should offer their help to carry out democratic reforms and function in a normal way as suits the total globalization era of systems taking into account the entropic equilibrium; otherwise these countries would not be able to exist and develop.

The above-mentioned four systemic laws are studied in detail in [22].

1.22. Conclusion

Universal systemic regularities in nature and society are, as a rule, of a limiting character; they let us study the influence of a society on the system of interference of the man and nature for the purpose of finding rational way of society development under conditions of limited resources.

The role of systemic regularities in human practice, first of all, consists in the fact that they have guiding influence on people's activity. Understanding causes as well as knowing the mechanisms of manifestation of system regularities makes it possible to avoid a lot of mistakes in the process of managing complex technical, economic, social, ecological, organizational and other systems. Taking into account these mistakes allows us to find the way out of a crisis. Environment plays a decisive role in the process of self-organization or in structure-forming processes.

If natural laws and regularities are violated or are not taken into account, and this is often the case, catastrophes and crises will follow just in the same way if aerodynamics laws are violated, the airplane will either not take off or fall to the ground. We should change our world outlook, our attitude to nature. We should see the ecological aspect or a new world outlook in any sphere of human activity – both technical and humanitarian.

Nature starts to degrade because of the human activity, which is too active and thoughtless; this is very dangerous, for instance, the concentration of carbon dioxide constantly increases, this endangers the existence of mankind; heavy showers which have flooded Europe are the consequence of thoughtless human activity. There if the only one way out – to correlate our actions with our real possibilities.

Basic processes in nature and society should, as a rule, proceed in accordance with universal laws and regularities and if the impact of external control on the system is made regardless of the regularities, it will not be effective. Therefore, basic systemic laws and regularities must nor be violated in the process of decision making. Managerial staff, executive managers, research workers, inventors should know universal systemic laws and regularities especially well. Universal systemic regularities determine systemic scientific thinking by making it more well-ordered.

Taking into account systemic regularities makes it possible to achieve integrity, stability, organization of technical, social, economic and organizational systems, to analyse complex systemic processes that take place in crisis situations and to plan the ways or the rational solution of management problems in different complex weakly formalized systems.

The regularity of entropic equilibrium makes it possible to determine the organization and disorganization of systems, choose methods of system management for the purpose of their improving and developing as well as understand initial causes of conflicts, wars, natural calamities and disasters. The regularity of entropic equilibrium in many respects determines conditions of self-organization or structurization and disorganization in systems of different kind. This regularity shows that global progress and prosperity will be impossible if people don't learn how to manage systems in such a way as to shift the excessive entropy to other systems, to the environment or outer space as well as to control the amplitude and frequency of entropic vibrations and the point of entropic equilibrium.

The regularity of system potential dependence on the system structure makes in possible to differentiate systems from pseudo-systems and to outline the rational organization of the interaction between the elements of a system.

The regularity in oscillatory and cyclic functioning of different systems enables us to distinguish cycles in the history of Russia from the year 862 to 1989. It is shown that this period is characterized by 375-year cycles during which Russia's disintegration into minor princedoms and later their amalgamation into one state occurred four times. On the basic of these cycles an attempt is made to prognosticate future historical processes of the XXI century.

Background general- systemic regularity allows to reveal "invisible" mobile and stationary objects in the system of radiolocation, optic detection, sonar systems, etc. and create new special systems and complexes for revealing "invisible" objects and protection devices, protecting airports, sea-ports and other major objects from terrorists, technogenic damages and disasters. Background regularity is very useful for research in physics, biology, medicine, etc.

On the basis of the existing objective natural systemic laws – 1) "unity and struggle of opposites"; 2) "critical point (moment) of crystallization", 3) "systemogenetic law"; 4) "island effect law"; 5) "dynamic entropic equilibrium", a concept for the system of *monitoring (control)* and management of basic elements of regions and a state can be elaborated.

REFERENCES

Alekseyev G. Energoentropics. M.: *"Znanie"*, 1983.

Astafiev O.N. Globalization as a Variant of Non-Linear Development of Socio-Cultural Processes/ Proceedings of the international conference "System Analysis at the turn of the Millennium". *Theory and practice.* Volume 1, M., 3-4 July, 2001.

Baranov V.M. Consciousness and Idealness. Idealness Problems in Science. *Proceedings of a conference.* M., 2000.

Bogomol S.I. Social System Potential / Proceedings of the international conference "System Analysis at the turn of the Millennium". *Theory and practice.* Volume 1, M., 3-4 July, 2001.

Bogdanov A.A. General Organizational Science (Tectology). L., *"Kniga",* 1925.

Grigoriev E.P. Reflexive Synthesis of Alternatives in the Metrics of "Golden Section" // *Reflexive Control.* 2001, No 2.

Kittel Ch. Statistical Thermodynamics. M., *"Nauka",* 1977.

Kobozev N.I. *Research in the Field of Thermodynamics, Processes, Information and Thinking.* M., Moscow Univ. Press, 1971.

Kuvichkin V.V. Manifestation of the Divine in the Physical World and in the Inner World / Proceedings of the international conference "System analysis at the turn of the XXI century". *Theory and practice.* Volume IY, M., 27-28 Februery, 2001.

Kurashvili B.P. Historical Logic of Stalinism. M., "Bilina", 1996.

Liyv E.H. Infodynamics, *General Entropy and Negentropy.* Tallin, 1998.

Lisitsin A.I. On the Axiomatics of Idealness. *Proceedings of a conference.* M., 2000.

Lopatin V.N. Security of Informational Technologies Economics under Conditions of Globalization // *Informatization Problems.* No 2, 2001.

Maslova N.V. *Noospheric Education.* M., 2002.

Medvedev U. Climate and earthquakes – the Main Thing in History // *"Vek"*, 2002, No 17.

Medvedev U. "The Inner Voice" Alerts. *"Vek"*, 2002, No 5.

Melkadze G. Regularity of Phenomena Reversibility // *Georgian Engineering News,* 2002, No5.

Muldashev E. Magic of "6666" // *Arguments and Facts,* 2001, No 23.

Muldashev E. When the Bottom Will Be Knocked out of the World // *Arguments and Facts,* 2001, No 21.

Novozhilov U.I. Where Is Russia Going? Science, Society, Man. *Proceedings of the Ural branch of Russian Academy of Sciences.* Issue 1. – Ekarerinburg, 2002.

Parkhomenko A. Cycles of Russian History. M., Agency *"Krechet"*, 2000.

Pranguishvili I.V. *Systemic Approach and General Systemic Regularities.* M., 2000.

Pranguishvili I.V., Abramova N.A., et al. *In Search of Approaches to Solving Problems.* M., 1999.

Pranguishvili I.V., Pashchenko F.F., Busiguin B. Systemic Laws and Regularities in Electrodynamics, Nature and Society. M., *"Nauka",* 2001.

Rife F. Berkeley's Course of Physics. Vol. 5, Statistical physics. M., *"Nauka",* 1986.

The Russian Elite Does not Fit the Times // *"Vek",* 2001, No 38.

Starovatova E., Askochenskaya A. Zone of Indeterminacy // *"Vremya MN"*, 25 May, 2002.

Kharitonov A.S. The Analysis of Sustainable Development Subject to Physical Specificity of Nature Functioning // Proceedings of the international conference "Moscow-Russia at the turn of the millennium", part II, M.: *"Informznanie"*, 2000.

Kharitonov A.S. Harmony and Symmetry – the Basis of the New Technology for Controlling Society. Problems of environment and natural resources. M., *"VNIITI"*, 2000, No 2.

Kharitonov A.S. Management According to "Golden Proportion" // Collected articles: Socio-economic and ecological aspects of sustainable development of cities. M., "Znanie", 1999.

Chirkov U. Scientific Environment // *"Literaturnaya gazeta"*, 2002, No 17.

Shapovalov V.I. Entropic World. Volgograd: *"Peremena"*, 1995.

Shapovalov V.I. Processes in Open Systems and Global Ecology // Herald of the regional fund for socio-political research and technology. *"STO"*. Collected reports: Vladimir regional administration, 1999, Issue 3.

Yuzvishin I.I. *Basics of Informology.* M., 2000.

Yanitsky I.N. Novelty in the Science about the Earth. M., *"Gelios"*, 2001.

Chapter 2

SYSTEMIC OPTIMIZATION OF DISTRIBUTED DECISION-MAKING SYSTEMS

2.1. INTRODUCTION

This chapter is dedicated to describing the tasks for optimizing distributed decision-making systems [10, 15, 33]. Regional development programs [6, 15, 24] are considered as an applied model. Such programs represent a complex of interrelated activities aiming at achieving certain development goals which are estimated according to social, economic and other criteria. The programs include a set of projects, and projects include sets of subprojects. Typical examples of such projects are projects for reforming and restructuring (PRR) enterprises [23], organizations, their subdivisions and other objects directed at improving the operating benefits of the object undergoing a reform. As a rule, each PRR helps to achieve simultaneously a number of goals and requires certain costs. Therefore, operating benefits may be estimated according to a number of parameters. Accordingly, the effect achieved as a result of PRR implementation is estimated according to several criteria. The management problem consists in choosing a great number of PRR-s to be implemented and supported in the first place under the input restraints.

In regional management the subject solving a management problem is represented by a team, i.e. it consists of several subjects having their own interests and preferences and different subjects can have different preferences. This can be exemplified by PRR-s, supported by the administration of the region where the enterprises undergoing reforms are located [23]. And different structural subdivisions of the regional administration may be interested in fulfilling different top-priority tasks. The fact that these interests may not always coincide is not indicative of the inefficiency of the control system, but reflects objectively existent complexity of the control object.

Organizational and social-economic systems having a number of administrative organs at one or several levels of hierarchy are called distributed control (DC) systems or distributed decision-making systems (DDMS) [19, 23]. The task of analyzing DDMS consists in describing a great number of equilibrium (in some respects) states and/or its participants' strategies. As for synthesis, its task consists in search for procedures promoting the interaction between decision-makers (DM) enabling them in concord to make effective management decisions.

Three problems are connected with DDMS study. Firstly, for determining the efficiency of this or that decision it is necessary to have a behavior model of a controlled subject depending on this decision (control). Secondly, all the participants of the system under consideration are active (i.e. they are able to make independent decisions – to choose conditions, to transmit information, etc. according to their own interests [12]. Therefore, both a controlled subject model and procedure promoting decision-makers' interaction should take account of this activity. Thirdly, the decision-making system apart from its state of distribution may have a complex multilevel structure. Therefore, the study of two-level models (PRR at the bottom level and "horizontal" DDMS at the top level) is not sufficient. It is necessary to study multilevel systems as well. In other words, in decision-making we observe both "horizontal" distribution (which is characterized by the existence of several decision-makers at the same level of hierarchy), and a "vertical" distribution in decision-making (which is characterized by the existence of several levels of hierarchy where decision-makers are located).

Goal-seeking behavior models are studied in the framework of such management theory schools in social-economic systems as: theory of active systems [12, 28, 32, 33], theory of hierarchical games [13, 14, 15], theory of choice [1, 19, 26] and others. Game-theoretical modeling is the main research method, which makes it possible to predict the behavior of participants in the system and choose the type of management which renders the system optimal from the point of view of the chosen criterion.

The basic model of the theory of active systems (AS) is a two-level static deterministic AS comprising a controlling organ – centre – at the top level of the hierarchy and one controlled subject – active element (AE) – at the bottom level. The results of game-theoretical analysis of the basic model are given in [14, 28, 30]. Multielement AS models [32], one-element AS with distributed control (DC) [17, 33] and multilevel AS are the closest to DDMS. Though multielement multilevel AS with DS have not been analyzed in literature (see classification systems and reviews in [32, 33]). Therefore, in this chapter first we analyze a separate class of AS models, and then the theoretical results are applied to constructing and analyzing SDMS for supporting PRR.

We begin with the description of the basic model – multielement two-level AS with DC for which we define centers equilibrium strategies and AE. Besides, we determine possibilities for coordinating AS participants' interests. Afterwards proceeding from the findings we study multilevel multielement AS with DC, for whom interests (including, criteria for estimating PRR) are harmonized in the framework of DDMS, which is modeled by cooperation of participants in the top levels of hierarchy.

2.2. Basic Model

Let us consider AS at whose bottom level we have n $\geq$ 1 AE, their strategy consists in choosing an action (the subscript numbers AE) $y_i \in A_i \subseteq \Re_+^{n_i}$, $n_i \geq 1$, $i \in I = \{1,2,...,n\}$ - to set AE. AE control is exercised by k centers (K = {1, 2,..., k} – a set of centers). The strategy of i-center consists in choosing a vector (the superscript numbers centers)

$u^i = (u_1^i, u_2^i, ..., u_n^i) \in U = \prod_{j \in I} U_j^i$, where $u_j^i \in U_j^i$ - controlling j-AE from i-center, $i \in K, j \in I$. The control vector for all the centers we designate $u = (u^1, u^2, ..., u^k) \in U = \prod_{i \in K} U^i$.

Preferences of each AE are reflected in a general case by a set of objective functions. The j- component for i-AE objective function we denote as $f_{ij}(\cdot), j = \overline{1, m_i}$ where m_i - "dimensionality of preferences"; a set of components of the AE objective function we define as $M_i = \{1,2,...,m_i\}$, $i \in I$. The objective function of i-AE depends on control $u_i = (u_i^1, u_i^2, ..., u_i^k) \in U_i = \prod_{j \in K} U_i^j$ and the operation vector $y = (y_1, y_2, ..., y_n) \in A'$, that is $f_{ij}(u_i, y) : U_i \times A' \to \Re^j$ where $A' = \prod_{i \in I} A_i$ - a set of efficient actions of all AE.

It should be noted that AE are strongly connected (according to the terminology offered in [30, 32]), that is the objective function of each of them in a general case evidently depends on actions of all the AE.

The preferences of the centers are reflected by their vector objective functions $\Phi^{ij}(u^i, y) : U \times A' \to R^j, j = \overline{1, q_i}$, where q_i - "dimensionality of preferences" of i-center, $i \in K$.

Let us denote a set of components of the i-center objective function $Q_i = (1,2,...,q_i), \mathrm{i} \in \mathrm{K}$.

With regard to objective functions and admissible sets we introduce the following assumption, to which we will abide by in the course of all our further reasoning unless special provisions are made.

A.1. Sets of all admissible strategies of all the participants in AS are compact, and objective functions are continuous for all variables.

Let us suppose that the order of functioning (i.e. the succession of information receipt and decision-making [28]) is as follows: first the centers simultaneously and independently (this assumption excludes from consideration the cooperative effects of the centers' interaction which is thoroughly analyzed in [17]) choose their strategies, which are functions of the future AE choices (this means that game Γ2 [14, 15, 18] is being considered) and report them to AE which then simultaneously and independently (this assumption excludes from consideration the cooperative effects of AE interaction) make their action decisions and thereby finally determine AS participants' benefits.

The peculiarity of the multielement AS together with DC under consideration is that two games are successively played in it [18, 33]; a game of centers (for determining controls) and AE game (for determining their actions), and the conditions of the latter game depend on the results of the former one. For the purpose of analysis equilibrium strategies of the centers in the first game are defined by the dependence of the AE game equilibrium on controls.

In the framework of the above mentioned DDMS peculiarity it is necessary to define concretely what will be understood by AS participants' rational behavior. Rational behavior is usually defined as follows [18]: in one-element AS it is considered that a decision-maker, whose preferences are described by scalar objective function, seeks by choosing parameters under his/her control to maximize the value of this objective function. If there are a few interacting subjects whose preferences are described by scalar objective functions, it is considered that they choose equilibrium Nash strategies. In the model under consideration there are several subjects whose preferences are reflected by their vector objective functions. Therefore, we work in the following assumption.

A.2. Under given controls AE choose strategies which, according to Nash, are equilibrium in the sense of disability by any of AS to improve simultaneously all the values of its vector objective function at the expense of one-sided deviation.

Formally assumption A.2 can be written down as follows: AE are found in one of the elements of the set *E(u)*, which is defined as:

$$E(u) = \{y^* \in A' / \forall i \in I \neg \exists y_i \in A_i : \forall j \in M_i \quad f_{ij}(u_i, y_i, y_{-i}^*) \geq f_{ij}(u_i, y_i^*, y_{-i}^*)$$

and $\exists l \in M_i : f_{ij}(u_i, y_i, y_{-i}^*) > f_{ij}(u_i, y_i^*, y_{-i}^*)\}$ where

$y_{-i} = (y_1, y_2, ..., y_{i-1}, y_{i+1}, ..., y_n)$ - the game situation for i AE,

$$y_{-i} \in A_{-i} \subseteq \prod_{j \neq i} A_j, i \in I. \quad (1)$$

In regard to the centers' behavior let us make the following assumption.

A.3. The solution to game of centers $E \subseteq U$, according to Pareto, is effective Nash equilibrium.

In [33] it is shown, that there are two modes of interaction between centers – a mode of cooperation (under which, according to Pareto, equilibrium is effective) and a mode of competition (under which centers "threaten" each other and equilibrium, according to Pareto, is ineffective). A total characteristic of equilibrium of centers' game in one-element AS together with DC is given in [17]. It should be noted that in assumption A.3. it is essential for the equilibrium of centers' game to be effective according to Pareto which meets the requirements of practice: below nonempty conditions of this set will be received and its dependence on the management of superior bodies in multilevel AS. will be analyzed.

Thus, the two-level AS model under consideration (generalization for mixed-level managements are considered below) is characterized by a number of AE n, a number of centers k, dimensionality of a permissible actions set AE n_i, dimensionality of preferences AE m_i and dimensionality of centers preferences q_i. In the model under consideration all these parameters can simultaneously adopt values of more than one. Therefore, particular cases of the model under consideration are (parameters of more than one are shown): basic control model in two-level AS ($l_i \geq 1$ [14, 28]), multielement AS with unitary control ($l_i \geq 1, n \geq 1$ [32]), and one-element AS together with DC($l_i \geq 1, k \geq 1, n_i \geq 1$ [14, 28]).

2.3. Properties of Optimal Controls

As the game under consideration is of Γ_2-type, the center strategy is the function of AE strategy, i.e. $u_i = u_i(y)$ (we will use the same symbols for designating the function and its value in case it does not cause confusion). We register free circumstances $y_{-i} \in A_{-i}$ and define the following values and sets (wherever maximum or minimum are used, it isassumed, that they are achieved).

A set of punishment strategies for i-AE according to j-component of its objective function:

$$U_{ij}^N(y_{-i}) = \{u_{ij}^N(y_{-i},\cdot) \in U_i / f_{ij}(u_{ij}^N(y_{-i},\cdot), y_i, y_{-i}) = \min_{u_i \in U_i} f_{ij}(u_i, y_i, y_{-i})\}, j \in M_i, i \in I \tag{2}$$

guaranteed value of j-component of the objective function of i-AE:

$$L_{ij}(y_{-i}) = \max_{y_i \in A_i} f_{ij}(u_{ij}^N(y_{-i}, y_i), y_i, y_{-i}),\ j \in M_i, i \in I \tag{3}$$

Work in the following assumption.

A.4. $\forall y_{-i} \in A_{-i}\ U_{ij}^N(y_{-i}) = U_{ij}^N,\ j \in M_i, i \in I$

A.5. $\bigcap_{j \in M_i} U_{ij}^N = U_{ij}^N \neq \varnothing,\ i \in I$.

A.6. $\forall y_i \in A_{-i}\ L_{ij}(y_{-i}) = L_{ij},\ j \in M_i, i \in I$.

Assumptions A.4-A.6 can be conventionally called "decomposition axioms", as they permit to decompose AE game and to implement coordinated and independent control of the components of their objective functions (see lemmas 2.1-2.2 and theorem 2.1 below). Semantically the meaning of these assumptions are as follows.

According to assumption A.4 a set of punishment strategies of any AE according to any component of its utility function does not depend on the circumstances of the game. It is carried out specifically if control is included in the AE objective function additively (see stimulation tasks below) or multiplicatively (in the latter case the second multiplier must be a constant sign).

Assumption A.5 means that for each AE there is such a set of controls which ensure punishment simultaneously in all the components of its utility function. This assumption is carried out specifically when only one control vector component is included in each objective function component, or when all the components of AE objective function have the same control monotony parts (decrease and increase), etc.

According to assumption A.6 the guaranteed value of the objective function of each AE in each component does not depend on the circumstances of the game. This property occurs, for instance, if assumption A.4 is carried out and monotony parts $f_{ij}(\cdot)$ do not depend on the circumstances (see below for assumptions concerning properties of the expenditure function and goals of stimulation).

The assumptions we worked in permit to receive a number of results characterizing the properties of optimal controls in the AS models under consideration.

Lemma 2.1. Let us assume that assumptions A.1-A.5 are carried out.

Register $(u,x) \in U \times A^{'} : x \in E(u)$ Define

$$u' = \begin{cases} u, y = x \\ u^N, y \neq x \end{cases} \tag{4}$$

where $u^N = (u_1^N, u_2^N, ..., u_n^N) \in U, u_j^N \in U_i^N, i \in I$. Then $x \in E(u')$.

Proof of lemma 2.1. Since $x \in E(u)$, then according to (1) we find: $\forall i \in I \neg \exists y_i \in A_i : \forall j \in M_i f_{ij}(u_i, y_i, x_{-i}) \geq f_{ij}(u_i, x_i, x_{-i})$ and $\exists l \in M_i : f_{ij}(u_i, y_i, x_{-i}) > f_{ij}(u_i, x_i, x_{-i})$. Let it be that $\exists i \in I$ and $\exists y_i \in A_i : \forall j \in M_i \ f_{ij}(u_i^*, y_i, x_{-i}) \geq f_{ij}(u_i^*, x_i, x_{-i})$ and $\exists l \in M_i : \ f_{ij}(u_i^*, y_i, x_{-i}) \geq f_{ij}(u_i^*, x_i, x_{-i})$.

If $y_i \neq x_i$, then the latter inequality contradicts (2), if $y_i = x_i$, then it should be solved as an equation.

Semantically lemma 2.1 means that if a certain vector of AE actions is a solution to AE game, then changing control so that it be different from punishment strategy only in case of choosing equilibrium strategies, does not change the equilibrium. Analogous results (corresponding to special cases of lemma 2.1) are given in [14, 17, 28, 30, 33]. It should be noted that, firstly, while passing from control *u* to control *u'* defined according to (4), benefits of centers do not change, and, secondly, assumption A.6 has not been used yet. It becomes essential for proving the following result.

Lemma 2.2. Let assumptions A.1-A.6 be carried out. Let us register $(u,x) \in U \times A' : x \in E(u)$.

Let us define

$$u_i^* = \begin{cases} u_i'(y_i, x_{-i}), y_i = x_i \\ u^N, y_i \neq x_i \end{cases} \tag{5}$$

where *u'* is defined (4). Then $x \in E(u^*)$. What is more, $x \in A^{'}$ - equilibrium in dominant strategies (EDS) of AE game.

Proof of lemma 2.2. The first statement of lemma is evident, therefore, let us prove that x – EDS, i.e. it occurs that

$$\forall i \in I \forall y_{-1} \in A_{-1} \neg \exists y_i \in A_i : \forall j \in M_i \qquad f_{ij}(u_i^*, y_i, y_{=i}) \geq f_{ij}(u_i^*, x_i, y_{-i}) \quad \text{and}$$
$$\exists l \in M_i : f_{ij}(u_i^*, y_i, y_{=i}) > f_{ij}(u_i^*, x_i, y_{=i}). \tag{6}$$

Let us assume that $\exists i \in I, \exists y_{-1i} \in A_{-i}$ and $\exists y_i \neq x_i, \forall j \in M_i f_{ij}(u_i^*, y_i, y_{=i}) \geq f_{ij}(u_i^*, x_i, y_{=i})$ and $\exists l \in M_i f_{ij}(u_i^*, y_i, y_{-i}) > f_{ij}(u_i^*, x_i, y_{-i})$.

Let us substitute (5), then under assumptions A.4 and A.5 we receive that $\forall j \in M_i L_{ij}(y_{ij}) \geq f_{ij}(u_i^*, x_i, y_{-i})$ and $\exists l \in M_i : L_{ij}(y_{-i}) > f_{ij}(u_i^*, x_i, y_{-i})$. Under A.6 the latter system of inequalities contradicts the punishment strategy definition (2).

The main result of lemma 2.2 consists in the fact that using control (5), the centers decompose AE game, i.e. make it beneficial (in the sense of Pareto-efficiency of corresponding benefits in utility function components) for each of them to choose actions x, aside from the game circumstances, that is aside from the choice of other AE. Analogous results (lemmas 2.2 corresponding to special cases) are given in [32]. It should be noted that in transition from control u to control u^*, defined according to (5), the benefits of the centers do not change.

Joint use of lemmas 2.1 and 2.2 makes it possible to state the following.

Theorem 2.1. Let the assumptions A.1-A.6 be substantiated. Then in the class of controls of type (5) an optimal control will be found.

Proof of theorem 2.1. Optimal control is a permissible control maximizing the efficiency criterion and representing equilibrium of the centers game (see assumption A.3) on condition that under this control AE choose equilibrium strategies (see assumption A.2). Let $u \in U$ be an optimal control. It guarantees some advantages to centers in equilibrium and forces AE to choose equilibrium actions. Consecutively using the results of lemmas 2.1 and 2.2 we build control u^* according to expressions (4) and (5) on control u. The solution to game AE will not change, centers benefits (and, consequently, the solution to their game) will not change either. Hence, u^* is an optimal control.

It should be noted that theorem 2.1 generalizes theorem 13 from [33] concerning a case of multielement AS, and theorem 4.4.1 from [32] concerning a case of AE vector preferences.

Let us define what will be understood by equilibrium of centers game. Let us assume that the dependence $y(u): U \to A'$, where $y(u) \in E(u)$, is known. This dependence can be determined by introducing conformity of choosing equilibrium $\psi(E(u)): Z^{A'} \to A'$ [17, 18, 28], that for each set of balances finds the only corresponding vector of actions which is equilibrium under given control. In other words, let us think that it is known which actions AE choose depending on controls (these actions are called actions carried out by given controls).

Let us according to assumption A.3 define equilibrium $E \subseteq U$ of centers game:

$$E = \{u^{*} \in U \setminus \forall i \in K \neg \exists u^{i} \in U : \forall j \in Q_i, \Phi^{ij}(u^{i}, u^{*-i}, y(u^{i}, u^{*-i})) \geq \Phi^{ij}(u^{*}, y(u^{*}))\} \text{ and } \exists l \in Q_i : \Phi^{il}(u^{i}, u^{*-i}, y(u^{i}, u^{*-i})) \geq \Phi^{il}(u^{*}, y(u^{*}))\}. \quad (7)$$

Expression (7) describes equilibrium of the centers game, i.e. makes it possible to analyze the properties of this equilibrium – its existence, etc. The task of synthesis - to give a constructive definition of nonempty conditions of this set, etc. – is solved below.

The efficiency of DDMS K^0 can be introduced as follows. Let the functional $\Phi^0(y): A' \to \Re^l$, be given. This functional reflects the efficiency of the state of the controlled system from the point of view of the metacenter (a controlling organ located at a higher level of the hierarchy than the centers, implementing direct control over AE). Semantically, Φ^0 reflects metacenter preferences regarding AE actions. Consequently, the DDMS efficiency is defined by the value of this functional for a set of AE actions carried out by equilibrium controls of AE actions.

As sets *E(u)* and *E* may comprise more than one element, it is necessary to more thoroughly define the state of AC. Let us introduce the following assumption reflecting benevolent attitude of AE and centers to the metacenter (under other equal conditions they will choose the most favourable strategies from the point of view of the metacenter, i.e. strategies maximizing the functional $\Phi^0(\cdot)$).

A.7. SDMS efficiency equals to

$$K^0 = \max_{u \in E} \max_{y \in E(u)} \Phi^0(y). \quad (8)$$

It should be noted that (8) defines not the control efficiency of the active elements by the centers, but DDMS efficiency, i.e. the totality of centers as a decision-making system. If we wished to define control efficiency, we should have calculated the maximum of a certain combination of objective functions of centers taking account of a set of solutions to the AE game and maximized this combination over a set of admissible or equilibrium controls. The aforesaid does not mean at all that functional $\Phi^0(\cdot)$ "has nothing to do" with AS under consideration: in case of one single center it may coincide with its objective function, then (8) will pass to the control efficiency criterion [28, 32]. Besides, this functional may be defined so that to maximalize the combination of AE utility functions (note that (1) does not guarantee the achievement of AE Pareto-effective in the sense of component objective functions of all the AE or the totality of components analyzed separately for each AE state).

Let us consider a special case of the model being described and, namely, the task of stimulation which is defined as a game Γ_2 [14], where there exist side payments [13] and the objective function of the first player does not quite obviously depend on control [33].

2.4. THE TASK OF STIMULATION

The following interpretations correspond to the task of stimulation. Each AE taking some actions incurs expenses. These expenses in a general case depend on the actions of all AE.

Control on the part of centers represents rewarding or punishing AE for the choice of certain actions (such control in a special case of carrying out the task of stimulation is designated by symbol $\sigma(\cdot)$. This means that the center (or centers in AS with DC) pays compensations to AE, which depend on the actions AE choose. The dependence of rewards on the actions is called a stimulation function (a stimulation mechanism, a stimulation system) which is additively included in the objective functions of the participants in AS - AE receives exactly the sum which is paid to him by the center and values it just as the center does (the difference of utility functions becomes essential in AS [8,30]).

Thus, a characteristic feature of the stimulation model is that reward is additively included in the objective functions of the participants in AS (with different signs: for the centers – with a minus, and for AE – with a plus) which makes it possible to speak about uniformity in defining the structure of objective functions: if there exists transferable commodity ("money") [18], the rest of the items included in the objective functions must be measured on the same scales and, consequently, must be comparable (see below for the definition of total expenses of centers on stimulation).

The task of the center is by applying this or that system of stimulation to impel AE to choose the most favorable actions. Numerous examples of stimulating AE are given in [13, 17, 30, 31, 32, 33]. Let us describe a formal model of stimulation in a multielement AS together with DC which is under consideration in this chapter.

Let us assume that centers preferences are scalar and that each center stimulates each AE by each component of the latter's objective function. Let us designate by $\{\sigma_{ij}^{l}(y)\}_{j\in M_i}$ - the totality of rewards to i i-AE from l-center, $v^l(y)=\sum_{i\in I}\sum_{j\in M_i}\sigma_{ij}^l(y)$ - total reward paid by l-center to all the AE, $\sigma_i^l(y)=\sum_{j\in M_i}\sigma_{ij}^l(y)$ - total reward, received by i-AE from l-center, $\sigma_{ij}(y)=\sum_{l\in K}\sigma_{ij}^l(y)$ total reward, received by i-AE from all centers, total reward by all the centers paid to i-AE according to j-component to its objective function, $j\in M_i, i\in I, l\in K$. Let us designate by $c_{ij}(y)$ j-component of i-AE expenses, $c_{ij}(y)=\sum_{j\in M_i}c_{ij}(y)$ - total expenses of i-AE, $i\in I, H^l(y)$ - the profit of l-center $l\in K$.

In the stimulation task the objective functions of the participants in the system look as follows:

$$\Phi^l(y,\sigma^l)=H^l(y)-v^l(y), i\in K, \tag{9}$$

$$f_{ij}(y,\sigma_i)=\sigma_{ij}(y)-c_{ij}(y), i\in I. \tag{10}$$

Let us bring semantic interpretations. Each AE chooses certain actions which in terms of PRR may be interpreted as efforts directed to implementing a certain project and bringing definite results. AE itself estimates its activity and other AE according to m_i criteria and the achievement of results $y \in A'$ from the point of view of j-criterion requires $c_{ij}(y)$ expenses from i-AE. Center with number $l \in K$ from reaching result $y \in A'$ gets profit $H^l(y)$ and pays rewards $\{\sigma_{ij}^l(y)\}_{j \in M_i, i \in I}$. Instances of applying such a description to PRR control are given below.

The centers' task is to choose such stimulation systems which would urge AE to take the most preferable steps from the centers' point of view. The backbone factor in the given model is the fact that the vector of the results of AE performance (actions) is common for all the centers which involves them in the game and forces them to coordinate their interests and arrange a settlement with each other (the process of coordinating interests and properties of compromise settlements are analyzed below in describing the centers game).

Let us work in the following assumption concerning objective functions and admissible sets of AC participants in the stimulation model under discussion.

A.8. Admissible sets $A_i \subseteq \Re_+^{m_i}$ are limited and include zero, $i \in I$.

A.9. 1) Function $c_{ij}(\cdot)$ is continuous in all the variables; 2) $\forall y_i \in A_i c_{ij}(y)$ does not decrease by y_i;. 3) $\forall y \in A' c_i(y) \geq 0$; 4) $\forall y_{-i} \in A_{-i} c_{ij}(0, y_{-i}) = 0, j \in M_i, i \in I$.

A.10. Stimulation functions are piecewise continuous and accept nonnegative values.

A.11. Centers' income functions are continuous in all the variables and reach maximum at nonzero actions of AE.

Lemma 2.3. If assumptions A.8-A.11 are substantiated, then assumptions A.1, A.4-A4-A.6 are substantiated.

Proof of Lemma 2.3. Assumption A.1 validity follows from A.8, A.9 and A.11. It follows from (2) and (3) that taking account of A.9 and A.10 we may choose punishment strategy $u_{ij}^N(y)$ identically equal to zero, aside from the circumstances, whence it follows that A.4 and A.5 are valid.

From A.8-A.10. and (3) it follows that aside from the circumstances (11) $L_{ij} = 0, j \in M_i, i \in I$, then assumption A.6 is valid.

Thus the introduced stimulation model is a special case of the above-mentioned general control model in multielement AS DC. Therefore, for the stimulation task lemmas 2.1 and 2.2 are valid as well as theorem 2.1 which means that each AE is rewarded only in case it chooses required actions aside from the choice made by other AE. Moreover, the following result is valid for the stimulation task.

Let us designate $\sigma = \{\sigma_{ij}^l(y)\}_{j \in M_i, i \in I, l \in K}$. We call the following solution to the task $\sum_{l \in K} \vartheta^l(x) \to \min_{\{\sigma_{ij}^l\}: x \in E(\sigma)}$ the centers' minimal total expenditure $v_{\min}(x)$ on stimulating the realization of operation vector $x \in A'$. (12)

Theorem 2.2. Let suppositions A.2, A.3, A.8-A.11 be substantiated. We fix arbitrary operation vector $x \in A'$. Stimulation system σ having the following property:

$$\sigma_{ij}(y) = \begin{cases} c_{ij}(y) + \delta_{ij}, y_i = x_i \\ 0, y_i \neq x_i \end{cases}, \tag{13}$$

where $\delta_{ij} > 0$ - arbitrarily small strictly positive constants, $j \in M_i, i \in I$, realize operation vector $x \in A'$ as the only EDS with centers' δ -minimal total expenditure on stimulation equal to:

$$v_{mm}(x) = \sum_{i \in I} \sum_{j \in M_i} (c_{ij}(x) + \delta_{ij}), \tag{14}$$

where $\delta = \sum_{i \in I} \sum_{j \in M_i} \delta_{ij}$.

Proof of theorem 2.2. The fact that $x \in A'$ is EDS of AE-game follows from lemmas 2.1-2.3. The fact that value (14) cannot be decreased is substantiated as follows. When applying the stimulation system having the property (13), in equilibrium the value of j-criterion of i-AE equals to δ_{ij}. The value δ can be made arbitrarily small. At the same time under assumptions A.8-A.10 any AE always (i.e. aside from controls) has an opportunity to get strictly zero utility by choosing zero action.

The availability of positive constants $\{\delta_{ij}\}$ is caused by the necessity of ensuring the uniqueness of EDS. In the framework of benevolence hypothesis they may be chosen as vanishing and operation vector $x \in A'$ will be realized as EDS with minimal (and not δ-minimal) total expenditure of centers on stimulation (see task (12)).

Consequence of theorem 2.2. Stimulation systems satisfying (13) are δ-optimal from the point of view of the centers' total stimulation expenditure.

In a special case, when there is only one center characterized by scalar preferences, theorem 2.2 changes to theorem 7 of work [33] which states that the optimal realized action will be the action ensuring maximum difference between the center's income function and AE expenditure function.

In semantic expression theorem 2.2 states that stimulation is "optimal" if it exactly and aside from circumstances compensates costs to each AE in case it chooses a required action and it equals to zero in case AE chooses any other action. This statement qualitatively may be considered synthesis of the following principles: compensation, decompensation and cooperation [35] for multielement AS with DC.

The results of theorem 2.1 make it possible to apply only control classes of types (4)-(5) determined by equilibrium (in centers game) control. Theorem 2.2 itemizes properties of optimal controls in stimulation task, but aside from properties (13), does not say which control centers are equilibrium. Therefore, we more thoroughly study the centers game.

2.5. The Centers Game

Let us first study a multielement two-level AS with DC in which centers have scalar preferences, and AE have vector preferences. In the stimulation task in the framework of benevolence hypothesis in accordance with the results of theorems 2.1 and 2.2 it is possible to restrict equilibrium (including Pareto-effective in accordance with A.3) strategies of the centers to those which are non-nil only in case AE chooses a required action.

We register $x \in A'$. Let us study the following centers strategies:

$$\sigma_{ij}^{l}(y) = \begin{cases} \lambda_{ij}^{l}, y_i = x_i \\ 0, \quad y_i \neq x_i \end{cases}, \ j \in M_i, i \in I, l \in K \tag{15}$$

where $\lambda_{ij}^{l} \geq 0$ - certain constants $j \in M_i, i \in I, l \in K$.

It follows from (14) that the following must take place:

$$\sum_{l \in K} \lambda_{ij}^{l} = c_{ij}(x) \quad j \in M_i, i \in I \tag{16}$$

Up to the end of this chapter we will consider assumptions A.1-A.11 to be substantiated.

Lemma 2.4. Stimulation system (15)-(16) realizes actions $x \in A'$ as EDS of AE game and is characterized by minimal expenditure on stimulation.

Proof of lemma 2.4. Stimulation system (15) taking account of (16) satisfies conditions (13) and (14). Consequently, according to theorem 2.2 in the framework of benevolence hypothesis $x \in A'$ - EDS. According to the consequence of theorem 2.2, stimulation expenditure in that case is minimal, and the stimulation system is optimal.

Consequence of lemma 2.4. When implementing stimulation task without efficiency loss it is possible to restrict the stimulation system classes to (15)-(16).

Designate

$$W^{l} = \max_{y \in A'} \{H^{l}(y) - \sum_{i \in I} \sum_{j \in M_i} c_{ij}(y)\} \tag{17}$$

$$\lambda^{l} = \sum_{i \in I} \sum_{j \in M_i} \lambda_{ij}^{l}, l \in K. \tag{18}$$

The value W^{l} shows the profit l-center may gain when managing AE staff (and paying rewards) alone (without other centers). Consequently, the condition of its individual rationality may be designated for it as a condition of taking (together with other centers) such AE actions that its profit will not be less than (17). In other words, it is profitable for l-center to take such actions $x \in A'$ whose profit subject to payments (18) is no less than (17):

$$H^l(x) - \lambda^l \geq W^l, l \in K \tag{19}$$

Bringing together (16)-(19), let us define the following set:

$$\Lambda(x) = \{\lambda_{ij}^l \geq 0, j \in M_i, i \in I, l \in K \setminus \tag{20}$$

$$H^l(x) - \sum_{i \in I} \sum_{j \in M_i} \lambda_{ij}^l \geq \max_{y \in A'} [H^l(y) - \sum_{i \in I} \sum_{j \in M_i} c_{ij}(y)], l \in K; \sum_{i \in K} \lambda_{ij}^l = c_{ij}(x), j \in M_i, i \in I\} \cdot$$

$\Lambda(x)$ set, which by analogy with [22] we call a region of compromise, characterizes a great number of such payment combinations on the part of the centers to active elements which are equilibrium of centers game.

The validity of the following result follows from theorems 2.1, 2.2 and lemma 2.4.

Theorem 2.3. Equilibrium of the centers game is control (15) which satisfies (20) and takes actions $x \in A'$ as EDC of AE game.

It follows from theorem 2.3 that realizable are such AE actions for whom corresponding sets are nonempty.

Let us designate

$$W = \max_{y \in A'} \{\sum_{l \in K} H^l(y) - \sum_{i \in I} \sum_{j \in M_i} c_{ij}(y)\} \tag{21}$$

$$x^* = \arg\max_{y \in A'} \{\sum_{l \in K} H^l(y) - \sum_{i \in I} \sum_{j \in M_i} c_{ij}(y)\} \tag{22}$$

Consequence of theorem 2.3. If $\exists x \in A'$: $\Lambda(x) \neq \varnothing$, then $\sum_{l \in K} W^l \leq W$.

The result of the consequence is substantiated by summing up expressions (19) of all the centers and by comparing of the amount received with (21) subject to set definition (20).

As it was noted in [33], the difference $W - \sum_{l \in K} W^l$ may be interpreted as an “integral” measure of coordination of centers’ interests.

The lemma given below makes it possible to aggregate the description of AE and represent them as one AE with expenditure function $c(\cdot): A' \to \Re^l$, which is defined as follows:

$$c(y) = \sum_{i \in I} \sum_{j \in M_i} c_{ij}(y). \tag{23}$$

Let us designate $\Lambda^* = \Lambda(x^*), \Lambda = \bigcup_{x \in A'} \Lambda(x)$,

$$\Lambda' = \{\lambda^l \geq 0, l \in K \setminus H^l(x) - \lambda^L \geq W^l, l \in K, \sum_{l\in K} \lambda^l = c(x)\}. \tag{24}$$

Lemma 2.5. For each $x \in A'$ the set $\Lambda'(x)$ is nonempty only when the set $\Lambda(x)$ is nonempty..

Proof of lemma 2.5. Register arbitrary $x \in A'$.If $\Lambda(x) \neq \varnothing$, then from (20) subject to (18) and (23) receive that $\Lambda'(x) \neq \varnothing$. Now let $\Lambda'(x) \neq \varnothing$. From (24) it follows that $\exists\{\lambda^l \geq 0\}$; $H^l(x) - \lambda^l \geq W^l, l \in K, \sum_{l\in K} \lambda' = c(x)$. Conditions $H^l(x) - \lambda^l \geq W^l, l \in K,$ in (20) and (24) coincide. The inequality system (18), $\sum_{l\in K} \lambda^l_{ij} = c_{ij}(x), j \in M_i, i \in I$, under (23) is always solvable (maybe the solution is nonunique) as regards $\{\lambda^l_{ij}\}$.

The statement of lemma 2.5 permits (including at the expense of theorem 2.3) to bring the given multielement AS with vector preferences AE to one-element AS DC with scalar preferences AE, which makes it possible to transfer the results received in [17, 33] to the class of models under consideration.

Lemma 2.6. If $\sum_{l\in K} W^l \leq W$, $\Lambda \neq \varnothing$.

Proof of lemma 2.6. The consequence of theorem 2.3 states that the condition $\sum_{l\in K} W^l \leq W$ is fulfilled if the set Λ is nonempty. In the given lemma a converse proposition must be proved, i.e. that if $\sum_{l\in K} W^l \leq W$ then $\Lambda \neq \varnothing$. Let us assume the contrary, i.e. let $\sum_{l\in K} W^l \leq W$ and $\Lambda = \varnothing$. Then according to lemma 2.5 $\Lambda' = \bigcup_{x\in A'} \Lambda'(x)$. Consequently, $\Lambda^{*\prime} = \varnothing$.

The latter condition may be written down as follows: $\forall\{\lambda^l \geq 0\}: H^l(x^*) - \lambda^L \geq W^l, l \in K$ is fulfilled $\sum_{l\in K} \lambda^l < c(x^*)$ (the sign of the latter inequality is caused by the fact that for the realizability of a corresponding action in (16) it is sufficient to require that $\sum_{l\in K} \lambda^l_{ij} \geq c_{ij}(x), j \in M_i, i \in I$.

Let us choose concrete $\{\lambda^l\}$ satisfying $H^l(x) - \lambda^l \geq W^l$, $l \in K$, notably, let us assume that $\lambda^l = H^l(x^*) - W^l$, $l \in K$. Let us sum up k of the latter equalities: $\sum_{l\in K} \lambda^l = H^l(x^*) - \sum_{l\in K} W^l$. According to the lemma conditions we receive that

$\sum_{l\in K}\lambda^l \geq \sum_{l\in K}H^l(x^*)-W$. Since according to definition $W=\sum_{l\in K}H^l(x^*)-c(x^*)$, $\sum_{l\in K}\lambda^l \geq c(x^*)$, which contradicts the assumption $\sum_{l\in K}\lambda^l < c(x^*)$.

The value (21) plays a considerable role in determining the set nonemptity $\Lambda(\cdot)$, namely, the following consequence of lemmas 2.5-2.7 include lemma 21 from [33] as a special case.

Consequence of lemmas 2.5-2.6. If $\Lambda^* = \varnothing$, $\Lambda = \varnothing$. If $\sum_{l\in K}W^l \leq W, \Lambda^* \neq \varnothing$.

Semantically, it follows from lemmas 2.4-2.6 that point $x^* \in A'$ is characteristic (typical) in the sense that if the set Λ is nonempty, then $\Lambda^* \subseteq \Lambda$, and vica versa, if $\Lambda^* = \varnothing$, $\Lambda = \varnothing$ then. This fact will be broadly used below when solving the task of coordinating the centers' interests, etc.

In conclusion of this paragraph we will briefly analyze a general case in which centers' preferences are vectorial. In the second part of this chapter it was stated that in a general case centers' preferences are reflected by their vectorial objective functions $\Phi^{ij}(u^l,y): U\times A'\to R^l, j=\overline{1,q_l}$, where q_i is "dimension of preferences" of i-center, $i\in K$. At the same time when regarding the stimulation task it was implied that centers' objective functions are scalar (see expression (9)): $\Phi^i(y,\sigma^i)=H^i(y)-v^i(y), i\in K$. All the results received in this part will remain valid if centers' preferences are vectorial and look as follows: $\Phi^{ij}(y,\sigma^i)=H^{ij}(y)-\alpha^{ij}v^i(y)$, $\alpha^{ij}\geq 0, \sum_{y\in Q_i}\alpha^{ij}=1, j=\overline{1,q_i}, i\in K$ and assumption A.3 is fulfilled in whose framework equilibrium of centers' game is defined by analogy with the equilibrium of AE game (see expression (1)). However, such a generalization is artificial; as it will be seen from further discussion, using scalar preferences of centers, defined on observation vectors of AE operation, appears to be sufficient for practical needs. It should be admitted that restricting centers' preferences to scalar preferences, we avoid the task of aggregated idea of multicriteria preferences, considering aggregates $H^i(\cdot), i\in K$ as specified (given) (if we have only a set of criteria and the aggregate is not defined, then for introducing it we can use methods described in [2, 3, 21, 34]).

Having described the centers' game, let us begin studying the issue concerning the coordination of their interests and analyzing the tasks of multilevel multielement AS distributed control.

2.6. Role of Top Management

Up to now we considered the two-level AC consisting of *n* AE on the lower level of the hierarchy and *k* centers on its upper level, the preferences of AE on a many-dimensional set of actions (the results of their activity) being vectorial.

In terms of PRR the model under consideration covers the following situations. There are *n* enterprises (or there are *n* sub-units at an enterprise) planning to realize PRR (for the sake of simplicity we may consider that every enterprise or sub-unit realizes only one PRR). The activity of every enterprise for realizing PRR is described by vector y_i of $m_i \geq 1$ length of its actions result (results of the activity – results of realizing the corresponding PRR). To achieve y_i results expenditures $c_i(y_i, y_{-i}) = \sum_{j \in M_i} c_{ij}(y_i, y_{-i})$ are needed which generally depend on the results of realizing other PRRs. In other words, non-separability of expenditures on AE reflects a possible interconnection and interdependence of PRR.

The distributed system of decision-making (e.g. regional administration or the leadership of an enterprise) includes k number of individual or collective subjects (centers); each of them estimates $H^l(y)$ effect from the realization of PRR set $l \in K$. The difference between effect values at different centers may be explained by differences in their criteria priorities assessment, the criteria that describe the results of PRR realization. Thus, for instance, one of the sub-units of regional administration may assume that social aspects of PRR results are the most important, another sub-unit may give preference to economic aspects, still another – to ecological ones, etc.

The situation which can be considered a special case of this model consists in the following: each PRR is described in terms of the same indexes i.e. $n_i = n'$, which n' – is the number of indexes reflecting the results of PRR, $i \in I$, and is estimated by AE according to the same criteria i.e. $M_i = M, i \in I$ and by centers – according to n' criteria, which are, however, the same for all the centers. In an even more special case $|M| = n'$. At the same time, in the stimulation task it is assumed that the effect and expenditures are measured in the same units and are included into criterion functions of AS participants additively; this results in the fact that effectiveness "maximization" in vectorial sense corresponds to the total effect maximization minus total expenditures.

Let's consider a model situation in which every project is described by two indexes (n'=2): increase in production volume y_{i1} and increase in quality y_{i2}; each of them is measured by a certain nonnegative real number, $i \in I$. Each AE estimates expenditure corresponding to vector $y_i = (y_{i1}, y_{i2})$ according to two criteria $(m_i = 2, i \in I)$ – change in fixed costs and change in variable costs; estimates of these costs are determined by expenditures $c_{ij}(y_{i1}, y_{i2}, y_{-i}), j = 1,2$. Thus, to get y_i result $c_i(y) = c_{i1}(y_{i1}, y_{i2}, y_{-i}) + c_{ij}(y_{i1}, y_{i2}, y_{-i})$ expenditures are necessary. Three centers (k=3) estimate y vector effect according to their own criteria $H^l(\cdot)$ – "social", "economic" and "ecological", the units of measurement being chosen with the purpose of defining total effectiveness which is defined as difference in total (according to all the criteria of the centers) effect and total (according to all the AE) expenditures.

Having completed semantic interpretations of the model description let us consider theoretical results of sections 4 and 5 of this chapter.

Theorem 2.3 and lemma 2.5 give the solution of management problem in terms of characterizing the set of "simple" decomposing types of management; the set contains an optimum solution and makes it possible to consider the aggregated description of agents within the limits of stimulation task (see criterion functions (9), (10)).

The results of section 5 give an answer to the question – in what cases the system of compensating expenditures of AE by the centers for the realization of PRR in possible; such a system according to which it is profitable for each AE to realize the corresponding PRR set as compared to the independent financing of any group of projects.

The results of lemmas 2.4–2.6 and their consequences characterize a set of actions of AE which can be realized, as well as a set of corresponding equilibrium strategies of the centers. In other words, for each AS these formal results help to find sets Λ and (25) $S = \{x \in A' \backslash \Lambda(x) \neq \varnothing\}$.

S set can be interpreted as a set of co-ordinated [9] (in the sense of co-ordinating interests and preferences of the centers in accordance with supposition A. 3) results of PRR realization i.e. a set of such vectors of AE actions that are profitable for the centers to realize jointly as Pareto-effective and individually rational equilibrium of their game. Nonemptiness of Λ set for a certain $x \in S$ - shows that it is profitable for centers to "chip in" to get this action realized. In accordance with lemmas 2.5–2.6 either $x^* \in S$ or $S = \varnothing$.

In this way, the results of section 5 give as the solution to the analysis problem (they answer the question – what types of control can be chosen and what results can be achieved in AE) i.e. the problems for determining the conditions for the existence of equilibriums and let each concrete AS build a set of equilibriums and examine its properties (nonemptiness and others). However, these results don't give the solution to the synthesis problem, i.e. they don't provide the possibility of getting an answer to the question as to which management must be chosen. The normative aspect gives rise to two more special questions:

- what is to be done if set Λ^* is empty (and if it is empty then set Λ is also empty – see corollary to lemmas 2.5 and 2.6)?
- if Λ^* set is not empty and contains more than one point, what kind of actual management must there be?

The expression "must be" needs to be explained. If we appeal to the fact, that it is better to choose the best management within the limits of rational behavior paradigm [18, 28], then the concept "the best" should be explained. As AS includes several active subjects the utility function of which makes it possible to determine the best alternatives from its point of view, then it is necessary to determine from whose viewpoint the best management should be chosen.

Several approaches are possible in this case. Within the limits of the normative theory of decision-making [1, 19, 26] it is possible to introduce the aggregated criterion which will comply with reasonable axioms and reflect the collective opinion of the centers. Then the problem will consist in realizing such actions and choosing such kinds of control which maximize the "collective" criterion. This is one of the possible approaches. However, as we are studying the problem of management, we will realize another approach (which includes the normative one).

Let's assume that AS has a metacenter or a set of metacenters located on a higher level of hierarchy, than the centers (through this assumption we pass from a two-level AS to a multilevel one). These metacenters, which we will call the higher leadership or "top management organs" are active i.e., first, they have their own interests and preferences for a great number of results of AE activity and center control and, secondly, they have the ability (and the right – within the limits of the organizational structure and the order of functioning under consideration) to appoint controls whose values are functions from the choice of players who make their moves later i.e. from the choice of centers and AE. For describing such a model the results of 2.2–2.5 sections can be used i.e. by analogy with 2.4–2.7 lemmas equilibrium strategies of metacenters and so on can be constructed; it is easy to see that the two-level model described above allows for the trivial increase in the number of hierarchy levels. We are not going to do it now because, firstly, in case of need the realization of this approach won't cause any difficulties, secondly, the description received will be too awkward to be used in practice. The following model is considered to be the alternative.

Let us assume that there is a metacenter, reflecting the interests of the system as a whole, where the operations researcher is located. In accordance with the results given in [17, 37] it is known that two modes of interaction between the centers of the sublevel of the hierarchy (in the three-level AS) – co-operation mode and competition mode.

Co-operation mode is characterized by the non-emptiness of Λ set as well as Pareto-effectiveness (from the centers' viewpoint) of their equilibrium strategies. In the competition mode centers have to "overpay" to AE and the corresponding equilibrium of their game is ineffective (in [17] it is proved that if transfers of utility and coalition forming are rarefied , the necessary condition for the centers of maximum coalition forming to be profitable is the non-emptiness of set Λ). It should be noted that in this chapter only the cooperation mode under consideration.

Therefore, if the interests of the metacenter fully coincide with the interests of centers (e.g. its criterion function is equal to the sum of those of the centres) then at empty Λ set the task of control consists in such an influence of the metacenter on subjects under control (the centers and AE), which would provide for the non-emptiness of Λ set. In other words, in this case the task of the metacenter (which we will call the co-ordination task see table 2.1) consists in shifting the centers from the competition mode to the co-operation mode.

Other ways of setting a task of control are also possible (see table 2.1).

Table 2.1. Harmonization and Control Tasks

	Metacenter does not have its own interests	**Metacenter has its own interests**
$\Lambda=\emptyset$	"Co-ordination task"	"Co-ordination task"
$\Lambda\neq\emptyset$	"Harmonization task"	"Control task"

Having discussed the role of higher organs of control and enumerated the possible tasks of the metacenter, let us pass on to accomplishing these tasks.

If the center does not have its own interests, Λ set in nonempty and consists of more than one point, the task of the metacenter is to harmonize the interests of the centers (harmonization task) i.e. to appoint a concrete point from Λ set on the basis of a certain mechanism (a decision-making procedure). If the metacenter has its own interests, then we will assume that at empty Λ set it, in the first place, is interested in shifting the centers to the cooperation mode (the above-mentioned co-ordination task), and then – in realizing AE actions and center control that are the most profitable for it (i.e. change of both Λ set and S set). The last problem is the management problem proper. Special cases of co-ordination and management problems are examined in [17, 33].

2.7. Co-ordination Task

In this section tasks of harmonizing interests of the centers by the metacenter not having its own interests (different from the interests of centers) except for the minimization of expenditures on harmonization are examined i.e. tasks of co-ordination (see section 2.6), consisting in realizing the co-operation mode.

Let $\Lambda = \varnothing$ and the task of the metacenter consists in shifting the centers from the competition mode to the co-operation mode. In compliance with lemmas 2.5 and 2.6 it is sufficient to ensure the nonemptiness of the set.

$$\Lambda^* = \{\lambda^i \geq 0, l \in K \setminus H^l(x^*) - \lambda^l \geq W^l, l \in K, \sum_{l \in K} \lambda^l = \sigma(x^*)\} \tag{26}$$

which will guarantee the accomplishment of $x^* \in \Lambda$.

According to lemma 2.5 for performing $\Lambda^* \neq \varnothing$ it is sufficient that $\Lambda^{*\prime} \neq \varnothing$ and for this, in its turn, according to the corollary of lemmas 2.5 and 2.6 it is sufficient that $\sum_{l \in K} W^l \leq W$.

Let's introduce the following values:

$$\chi_l^* = \max\{H^l(x^*) - W^l; 0\}, l \in K \tag{27}$$

$$\sigma_l^* = \max\{W^l - H^l(x^*); 0\}, l \in K \tag{28}$$

$$\Phi^0 = c(x^*) + \sum_{l \in K} (\sigma_l^* - \chi_l^*). \tag{29}$$

Let the control consist in imposing fines on the center $\{\chi_l^*\}$ and paying remuneration $\{\sigma_l^*\}$. Then, in accordance with (9) the criterion function of l-center with regard to the control on the part of the metacenter will be:

$$\Phi^l(y, \sigma^l) = H^l(y) - v^l(y) - \chi_l^* + \sigma_l^*, l \in K. \tag{30}$$

Semantically, (27) determines maximum values of the summary stimulation, which can be paid by l-center without violating the conditions of its individual rationality in the game of centers, (28) are minimum values that l-center should be paid in addition so that conditions of its individual rationality in the game of centers are fulfilled. Payments of AE centers may be hypothetically replaced by their interaction with the metacenter which collects the taxes from the centers (27) and pays compensation (28) to the centers and $c(x^*)$ – elements (see expression (29), reflecting the criterion function of the metacenter).

It should be noted, that semantic analogies can be drawn between mechanism (27)–(30) and the mechanism of key agents, sometimes called Growes mechanism [38, 39, 40]. (see also survey [8] and results in [27, 37].

It is easy to check that Λ^* set is not empty, as the metacenter compensates for expenditures of AE, and the inequalities of individual rationality in all the centers are fulfilled as equalities.

Lemma 2.7. $\Phi^0 = \sum_{l \in K} W^l - W$.

Lemma 2.7 proving. In accordance with (29)

$$\Phi^* = c(x^*) + \sum_{l \in K}(\sigma_l^* - \chi_l^*) = c(x^*) + \sum_{l \in K} \max(W^l - H^l(x^*);0) - \sum_{i \in K} \max(H^l(x^*) - W^l;0) =$$
$$= c(x^*) + \sum_{l \in K} W^l - \sum_{l \in K} H^l(x^*) = \sum_{l \in K} W^l - W$$

Semantically, in accordance with 2.7 lemma Φ^0 value may be considered as a minimum value of the own means of the center; this value is necessary for ensuring the co-operation between the centers.

Lets us define the following dependencies:

$$\chi^l(x) = \max\{H^l(x) - W^l;0\}.l \in K. \tag{31}$$

$$\sigma^l(x) = \max\{W^l - H^l(x);0\}.l \in K. \tag{32}$$

$$\Phi^0(x) = c(x) + \sum_{l \in K}(\sigma^l(x) - \chi^l(x)).x \in A'. \tag{33}$$

Lemma 2.8. $\forall x \in A' \quad \Phi^0(x) \geq \Phi^0$.

Lemma 2.8 proving. Let's assume that $\exists x \in A' : \Phi^0(x) < \Phi^0$. By substituting (27–29) and (31)–(33) we get: $\sum_{l \in K} H^l(x^*) - \sum_{l \in K} H^l(x) < c(x) - c(x^*)$, which is contrary to the definition of value x^*. •

Semantically, lemma 2.8 means that the realization of action x^* in case of competition mode needs the minimum expenditure and in case of co-operation – is the most profitable.

The following theorem is the immediate corollary of lemmas 2.7 and 2.8 (see also theorem 24, which considered a special case, in [33]).

Theorem 2.4.

The realization of operation x^* is optimal for a metacenter not having its own interests. Its expenditures on harmonization (29) comply with the following correlations: if $\sum_{l \in K} W^l \leq W$. then, $\Phi^0 \leq 0$; if $\sum_{l \in K} W^l \geq W$, then $\Phi^0 \leq 0$.

In accordance with the semantic interpretation given above, if $\sum_{l \in K} W^l \leq W$ then the co-operation regime between centers is realized without the interference of the metacenter, which, according to (27)–(29) has a possibility of obtaining a non-negative value Φ^0. If, on the other hand, we have $\sum_{l \in K} W^l \geq W$, then the realization of the co-operation mode between the centers requires the interference of the metacenter, which, according to (27)–(29), spends its own means (the negative sign is conditioned by this) again at Φ^0 rate.

The result of 2.4 theorem makes it possible to generalize the financing procedure in case there are several levels [25] of controlling organs in a multi-level system: for metacenters of the third level (from the bottom) of the hierarchy a region of compromises can be introduced in the definition of which it should be required that their total payments be equal to Φ^0. It this region is empty, controlling organs of higher levels should provide for the fulfillment of the conditions of the individual rationality of metacenters. By analogy with this, compromise regions are determined for the higher levels of hierarchy.

Thus, theorem 2.4 gives the solution to the co-ordination problem in case when the metacenter does not have its own interests. Let us consider the problem of harmonization of center interests (see section 2.6) when Λ set is not empty.

2.8. Harmonization Problem as the Problem of Resource Distribution

In accordance with the results of section 2.7, Φ^0 value characterizes the degree of harmonization of centers interest i.e. the sum that has to be paid in addition or can be withdrawn (depending on the sign of the expression (29)). If set Λ is empty, the optimum control of the metacenter is determined in theorem 2.4 and lemmas 2.7 and 2.8. Therefore, let us examine the case when set Λ is not empty. Then, according to the corollary to 2.5 and 2.6 lemmas, set Λ^* is not empty either. Owing to theorem 2.4, it is the most profitable for the metacenter to realize operation x*. If set Λ^* consists of only one point, then, obviously, Φ^0=0. If set Λ^* has more than one point, then the task of the metacenter may consist in determining the concrete realizations of control.

One of the variants is to withdraw all the surpluse s of the centers according to (27)–(30) – then the payments of the centers are plain. The alternative is the redistribution of the surpluses between the centers, i.e. the choice of a specific point from set Λ^* will occur in accordance with a certain procedure (decision-making mechanism), established by the metacenter and not due to independent agreement between the centers. This approach is advantageous because the centers do not have to calculate the equilibrium (which can prove important if their information resources are limited).

The mechanisms of resource distribution [5, 11, 27] present a large class of decision-making procedures in managing organizational (active) systems and are studied in the theory of active systems [12], the theory of hierarchical games [14, 15], the decision-making theory [19, 26, 27] and other sections of the theory of socio-economic systems management. Mechanisms of income or expenditures distribution constitute their special case [4, 27].

Resource-distributing mechanisms in multi-level systems were studied in [25], and, from the point of view of their activity, in [11, 29].

Let us denote

$$\mu^l = H^l(x^*) - W^l, l \in K. \tag{34}$$

If the metacenter uses mechanism (27)–(30) in equilibrium, the values of criterion functions of the centers are $W^l, l \in K$ and their payments (taking into account fines and incentives by the metacenter) equal to $\mu^l, l \in K$. These payments realize action x^* and are Pareto-optimal. Consequently, the metacenter has to distribute resource Φ^0 between k centers. Let us denote by g^l -the amount of the resource allotted by the metacenter for the l-center, then $R = -\Phi^0 \geq 0$ owing to the above-mentioned assumption, which is considered as $\Lambda \neq \emptyset$. If the metacenter has not got its own interests, then, obviously, budget limit must be introduced.

$$\sum_{l \in K} g^l = R. \tag{35}$$

Let us note that if the metacenter has its own interests, then it can distribute value $(l - \xi)R$), between the centers where $\xi \in [0;1]$ can be interpreted as the "tax" rate that is paid by the centers to the metacenter.

If the information is well distributed i.e. if all the participants of AS are well-informed about all criterion functions and feasible sets, then centers cannot influence the additional profit rate $g^l$of each center (as compared to W^l). Therefore, the resource–distributing mechanism (by which we, in this case, will mean the principle of value determination $\{g^l \geq 0\}$ that complies with [35]) may be set in different ways. Let us consider some of them that are widely used in practice [11, 23] and have the most transparent semantic interpretation.

1. *Principle of equal distribution* ($\forall l \in K \; g' = \text{Const}$):

   $g' = R / k, l \in K$.

   Evidently, while using the principle of equal distribution (35) is fulfilled and $g' \geq 0, l \in K$.

2. *Priority principle* ($\forall l \in K\ g^l / \gamma^l =$ Const, where $\{\gamma^l > 0\}$ - are constants, reflecting the priorities of the centers from the point of view of the metacenter, $\sum_{l \in K} \gamma^l = K$):

$$g^l = \gamma^l R / k, l \in K.$$

While using the priority principle, (35) is fulfilled and $g' \geq 0, l \in K$. At equal priorities, the priority principle of resource distribution passes into the principle of equal distribution.

3. P*rinciple of equal profit* ($\forall l \in K\ W^l + g^l =$ Const):

$$g' = \frac{R + \sum_{l \in K} W^l}{k} - W^l,\ l \in K\ .$$

While using the principle of equal profit, it is obvious, that (35) is fulfilled, but to fulfill $g^l \geq 0, l \in K$ it is necessary to demand $W^l \leq W / k$, which is a much stronger condition than condition $\sum_{l \in K} W^l \leq W$ of non-emptiness in the compromise area.

4. *Principle of equal profitability* ($\forall l \in K\ (W^l + g^l) / \mu^l =$ Const):

$$g^l = W \frac{\mu^l}{\sum_{l \in K} \mu^l} - W^l, l \in K.$$

While using the principle of equal profitability it is evident that (35) is fulfilled, but to fulfill $g^l \geq 0, l \in K$ it is necessary to demand that $\mu^l / W^l \geq \sum_{l \in K} \mu^l / W, l \in K$ take place.

5. *Principle of proportional investment* ($\forall l \in K\ g^l / \mu^l =$ Const):

$$g^l = R \frac{\mu^l}{\sum_{l \in K} \mu^l}, l \in K.$$

While using the principle of proportional investment, (35) is, evidently, fulfilled, but to fulfil $g^l \geq 0, l \in K$ it is necessary to demand that $H^l(x^*) \geq W^l, l \in K$ take place.

The enumeration of different mechanisms of resource distribution (principle of determining the profits that are paid by the metacenter) could be continued (e.g. on the basis of the proportionality of the resource to values $\{\chi_l^*\}, \{\sigma_l^*\}$ and their combinations) by using the above-used methods. But we will examine the mechanisms of resource distribution under conditions of incomplete information i.e. when there exist undefined parameters (unknown to the decision-maker – DM). We will consider the metacenter as a DM, and those characteristics of the functions center expenditures that are unknown to the metacenter will serve as parameters (AE characteristics will be considered as known for certain).

Thus, let $H^l(y,r^l)$ - be the l-center income function that depends on the action vector $y \in A'$ and parameter $r^l \in \Omega^l$ the type of l center that is unknown to the metacenter. Let us denote $r = (r^1, r^2, ..., r^k)$ as the vector of center types, $r \in \Omega = \prod_{l \in K} \Omega^l$ . .

It follows that in the model under consideration all the values depend on the type of the center:

$$x^*(r) = \arg\max_{x \in A'}[\sum_{l \in K} H^l(x,r^l) - c(x)], \tag{36}$$

$$W^l(r^l) = \max_{x \in A'} H^l(x,r^l) - c(x), l \in K, \tag{37}$$

$$\mu^l(r) = H^l(x^*(r) - W^l(r^l), l \in K, \tag{38}$$

$$\Lambda^*{}'(r) = \{\lambda^l \geq 0, l \in K \setminus H^l(x^*(r), r^l) - \lambda^l \geq W^l(r^l), l \in K, \sum_{l \in K} \lambda^l = c(x^*(r))\}, \tag{39}$$

$$R(r) = W(r) - \sum_{l \in K} W^l(r^l). \tag{40}$$

It is interesting to note that in this mechanism of resource distribution, unlike all the known mechanisms [4, 5, 11, 27, 29, 36], the amount of the distributed resource obviously depends of the center type.

Let us denote by $\Omega^* = \{r \in \Omega \setminus \Lambda^*{}'(r) \neq \varnothing\} \subseteq \Omega$ -the set of such vectors of center types, when co-operation mode is possible.

As the exact values of center types may be unknown to the metacenter, let us consider the mechanism of communicating information [8]; according to this mechanism centers inform the metacenter about the estimates of their types and the metacenter uses these estimates to determine the amount of the resources given to this or that AE (usually, the rule that brings the plans in correspondence with the communication vector is called the planning procedure [28].

The problem of manipulation is characteristic of planning mechanisms (to be more precise-mechanisms of communicating information); if the information communicated to the metacenter cannot be verified by it, the centers will communicate the information which will result in the most advantageous decisions for the center (e.g. if Λ set is empty, it is profitable for the centers that the metacenter would compensate to the agent for as many expenditures as possible). Therefore, it is very important to perform the task of studying manipulation mechanisms, in particular, determining a set of non-manipulated mechanisms for resource distribution i.e. such mechanisms in which communicating reliable information is profitable for AE. Let us formalize the last statement.

Let us consider the l-center preference function which reflects the dependence of its profit on the communication vector $s=(s^1,s^2,...,s^k)\in S$ of all centers (including its communication $s^l\in S^l, l\in K, S=\prod_{l\in K} S^l$):

$$\varphi^l(s,r^l)=H^l(x^\bullet(s),r^l)-\mu^l(s)+g^l(s), l\in K \tag{41}$$

where $\{g^l(\cdot)\}^{l\in K}$ is the resource-distribution procedure.

Within the limits of theoretical-game models [18] it is considered that under conditions of game uncertainty rational behavior of the subjects is modeled by the supposition that they will be in certain equilibrium. Therefore, the profitability of communicating reliable information means that reliable communications are in equilibrium [18]. Let us define the equilibrium in dominant strategies (EDS) $s^d\in S$ in the following way. Strategy $s^{ld}\in S^l$ is the dominant for l-center if

$$\varphi^l(s^{ld},s^{-l},r^l)\ge\varphi^l(s^l,s^{-l},r^l)\forall s^l\in S^l,\forall s^{-l}\in S^l \tag{42}$$

where $s^{-l}=(s^l,...,s^{l-l},s^{l+l},...,s^k)\in S^{-l}=\prod_{j\ne l} S^j$ - is the game setting for l-center. If each center possesses a dominant strategy, then the corresponding vector is called EDS.

Nash's equilibrium is vector $s^N\in S$, this is the kind of vector, that

$$\varphi^l(s^{lN},s^{-lN},r^l)\ge\varphi^l(s^l,s^{-lN},r^l)\forall s^l\in S^l,\forall l\in K. \tag{43}$$

Let us denote EDS set as $E^d(r)$ and the set of Nash's equilibriums as $E^N(r)$. Let us call the mechanism in which $r\in E^d(r)$ or $r\in E^N(r)$ non-manipulated (it is obvious, that $E^d(r)\subseteq E^N(r)$).

Let us consider the following model. Let us denote the vector of AE actions as $x_i^*\in A'$ - this vector can be most profitably realized by the collective of $K\setminus\{i\}$ centers:

$$x_i^*=\arg\max_{x\in A'}[\sum_{l\in K\setminus\{i\}} H^l(x)-c(x)], i\in K, \tag{44}$$

$$\Lambda_i^*=\{\lambda^l\ge 0, l\in K\setminus\{i\}\setminus H^l(x_i^*)-\lambda^l\ge W^l, l\in K\setminus\{i\}, \sum_{l\in K\setminus\{i\}}\lambda^l=c(x_i^*)\}, i\in K, \tag{45}$$

a set of co-ordinated payments of the centers from set $K\setminus\{i\}$ at the realization of x_i^* vector of AE actions. Total utility of these centers (without i-center) equals to

$\sum_{l \in K \setminus \{i\}} H^l(x_i^*) - c(x_i^*)$. If i-center is present, the total utility of this set of centers does not exceed (the maximum is achieved when the condition of individual rationality is an essential condition for i-center) $\sum_{l \in K \setminus \{i\}} H^l(x^*) - c(x^*) + H^i(x_i^*) - W^i = W - W^i, i \in K$.

Value $\delta^i = W - W^i - \sum_{l \in K \setminus \{i\}} H^l(x_i^*) - c(x_i^*)$, which can be interpreted as an excess [18], characterizes the change in the utility of all the centers except for the fixed one, while adding it to AS. Therefore, payments from the metacenter may be based on $\{\delta\}_{l \in K}$ values (see also the mechanism of key agents, according to which each participant is taxed; the tax is equal to those losses which other participants of AS suffer owing to his presence [27, 37, 38]). Payments by the metacenter can also be based on the $W - W^l$ difference that characterizes the maximum utility, which other centers can get owing to the presence of l-center. If the resource, obtained by each center is proportional to this investment, then the corresponding principle of resource distribution may be written as follows:

$$g'(s) = (W(s) - \sum_{l \in K} W^l(s^l)) \frac{W - W^i}{kW - \sum_{l \in K} W^l}, i \in K. \tag{46}$$

It is obvious, that mechanism (46) is anonymous and balanced, but, in general, it can be manipulated.

Let us consider another mechanism of resource distribution:

$$g'(s) = (W(s)/k - W^i(s'), i \in K. \tag{47}$$

It is also anonymous and balanced and, as in the previous case, manipulated.

The analysis of information distortion by the centers in mechanism (47) prompts that to avoid manipulation, complete redistribution of utility among centers should be achieved; this is possible within the limits of the following procedure: the mechanism should be of the type that

$$\varphi'(s) = [\sum_{l \in K} H^l(x^*(s), r^l) - c(x^*(s))]/k, i \in K \text{ takes place.} \tag{48}$$

Semantically, within the limits of mechanism (48) the metacenter uses institutional control and instructs the centers to forward the estimates of their income functions to the metacenter; on their basis the most profitable AE action can be calculated and the actual utility can be equalized (to achieve this the metacenter should, for certain, observe these utilities – in expression (48) the true values of center types are presented in income functions).

Let us assume that $H^l(y,r^l)$ functions are continuous and monotonic according to r^l parameters, and Ω^l sets are compact, $l \in K$. Then the following result holds true.

Theorem 2.5. Mechanism (48) is anonymous, effective and non-manipulated, but in general case it does not fulfil the conditions for the individual rationality of centers.

Theorem 2.5 proving. The anonymity of mechanism (48) is evident, as it is symmetrical with regard to the rearrangement of the agents.

Let us assume that all the centers except for i center, have conveyed reliable information i.e. $s^l = r^l, l \in K \setminus \{i\}$. The utility of the i-center equals to:

$$\frac{1}{k}[\sum_{l \in K} H^l(x^*(s^i,r^{-i}),r^l) - c(x^*(s^i,r^{-i}))].$$

On account of (36) within the limits of the assumptions introduced, the maximum of this expression is achieved at $s^i = r^i$. From (43) is follows that imparting the right information by the centers is Nash's equilibrium, i.e. the mechanism is non-manipulated. In addition, it is evident that the mechanism under consideration brings each center to the same level of utility $W(r) \setminus K$. And, finally, the effectivenesss of Pareto mechanism (48) results from the fact that is maximizes the total utility of the centers at conveying reliable estimates by them.

The unbalanced state of mechanism (48) and the possibility of violating the conditions of the individual rationality are characteristic features of non-manipulated mechanisms with transferable utility (see survey in [27]).

Example 2.1. Let $H^l(y,r^l) = 2r^l\sqrt{y},\ l = 1,2,\ c(y) = y,\ A = \Re_1^+$.

Let us calculate the following values in accordance with (36)–(40): $W^l(r^l) = (r^l)^2, x^*(r) = (r^l + r^2)^2, W(r) = (r^l + r^2)^2, \mu^l(r) = r^l(r^l + 2r^{-l})$.

If $\Omega = [0;\infty)^2$, then $\forall r \in Q\ W^1(r^1) + W^2(r^2) \le W(r)$, consequently, $\Omega^* \ne \varnothing$.

The metacenter can distribute

$$R(r) = \mu^1(r) + \mu^2(r) - c(x^*(r)) = W(r) - W^1(r) - W^2(r) = 2r^1r^2$$

units of the resource among the centers.

Let us consider the deviation of the first center from Nash's equilibrium. Within the limits of the mechanism (48) we calculate: $x^*(s^1,r^2) = (s^1 + r^2)^2, \varphi^1(s^1,r^2) = (r^1 + r^2)(s^1 + r^2) - (s^1 + r^2)^2/2$. Maximum of the last expression according to s^l is reached at $s^l = r^l$.

Thus, the considered resource distribution mechanisms make it possible to solve the problem of harmonizing center interests in the sense of sampling of a concrete variant from the set of Pareto-optimal variants. The advantage of the suggested approach consists in the fact that the initial task – determining the equilibrium payments of the centers i.e. a specific point of Λ set (if it is nonempty) seems to be more difficult to analyze as this set may be specified in a complex manner and have a complex configuration. The use of mechanism (48) according to theorem 2.5 makes it profitable for each center to convey reliable information; this makes it possible to achieve effective distribution of the resource among the centers under conditions of incomplete information offered to the metacenter.

Up to now we have been considering co-ordination and harmonization problems that correspond to the case when the metacenter does not have its own interests (see table 2.1). Let us go over to analyzing the case when the metacentre has its own interests i.e. to studying the problem of control.

2.9. Management Problem

Let us assume that the metacenter has its own interests that are different from those of the centers. The interests of the metacenter on the *A'* set of vectors of agents' actions will be described by the income function $H^0(y): A' \to \Re^1$, which we will consider continuous. Special cases of control problem, in which the metacenter imposes an income or profit tax on the centers, are studied in [17].

The role of the metacenter may possibly be interpreted by adding it to the set of centers. The set of trade-offs will be changed (apriori it is unknown whether it will grow or become smaller), equilibrium payments of the centers, etc. will be changed. However, this case more corresponds to the qualitative increase in the composition of centers than to introducing the new level of control. As it was noted in [14, 33], the presence of the metacenter implies the presence of the player. Who has the right to set the rules of the game for other players, to choose his own strategies as functions from the choice of other players. Therefore, let us consider the case when the metacenter has the right to make "the first move" and establish control over the centers; later, the centers choose payments to active elements.

The results of the previous statement have prepared everything that is necessary for setting and solving the problem of control. Expression (33) determines the budgetary balance of the metacenter according to the action realized. Therefore, the goal function of the metacentre $\Phi(x)$ may be defined as the difference between its income and expenditures (33):

$$\Phi(x) = H(x) - c(x) + \sum_{l \in K} (\chi^l(x) - \sigma^l(x)), x \in A'. \tag{49}$$

Let us adduce meaningful interpretations. In monograph [31] mechanisms of multi-level hierarchical organizational systems were studied. In particular, factors influencing the management efficiency in such systems were examined. One of the factors was the so-called "economic" factor which consists in the change of both financial, material and other resources of the system and the change of the participants of the system (controllable subjects, middle management organs and so on), possessing their own interests.

The change in the control effectiveness owing to the introduction or consumption of resources in the process of changing the composition of elements of the organizational system takes place in two-level systems as well. For instance, introduction of a new controlled object may broaden the possibilities of a system and, at the same time, increase expenditures for maintaining its activity. In other words, in general case, the economic factor reflects the balance of resources (conventionally incomes and expenditures) in the tasks of forming the system structure. Thus, for instance, introduction of a new sublevel in the hierarchy of the organization may, on the one hand, improve the co-ordination of the subordinates but, on the other hand, it may need extra costs for new managerial personnel. At the same time the introduction of new additional levels of management can only deteriorate the co-ordination of the staff activity, for example, by delaying the decision-making process. In the model under consideration (DDMS) the economic factor shows in the fact that the centers, drawing their own income from AE activity, take upon themselves part of the expenses connected with AE stimulation.

Theorem 2.6.

Mechanism (31)–(33) is optimal; within its limits operation

$$x^0 = \arg\max_{y \in A;} \Phi(y) \tag{50}$$

is realized.

Proof for theorem 2.6. Let us assume that there is another kind of control that is different from

$\{\chi^l(x), \sigma^l(x)\}$ which realizes operation $x' \neq x^0$ so that

$$H^0(x') - v(x') > \Phi(x^0) \tag{51}$$

where $v(\cdot)$ - are minimum expenditures of the metacenter for the realization of the corresponding operation. On account of lemmas 2.5–2.7 conditions of realizability include nonemptiness of set (26) which in its turn, is set by a system of inequalities $H^l(x') - \lambda^l \geq W^l, l \in K$.

$$\Phi^l(y^l) = H^l(y) - v^l(y) - \chi^l(y) + \sigma^l(y), l \in K. \tag{52}$$

By virtue of theorem 2.3, stimulation of the centers in realizing the operation is minimum. Proceeding from (52) and from the fact that in (50) the maximum for all the AE operations is calculated, we get a contradiction with (51).

The qualitative substantiation of the result of theorem 2.6 consists in the following: as at using controls (31)-(33) utilities of all AS participants, except for the metacenter, are equal to reserve utilities and in (50) the optimal operation realized with minimum expenditures of the metacenter on stimulation is determined, then $\Phi(x^0)$ is the absolute optimum of efficiency criterion.

It should be noted that by combining the results of this and the previous sections, requirements for providing all the AS participants with certain guaranteed utility levels (reserve utility) may be introduces in the control mechanism. Reserve utility will additively be included in conditions of individual rationality, so the compromise set will get narrower along with the increase in reserve utilities.

The result of theorem 2.6 can be directly generalized in case of four and multi-dimensional DDMS, as well as systems with distributed control on all levels of management.

2.10. Conclusion

In this chapter game-theory and optimization models of distributed decision-making systems have been studied; for this purpose equilibrium strategies of the management organs and controlled subjects have been examined; tasks of both harmonizing the interests of DDMS participants and managing their interaction have been solved.

The long-term direction of future research may consist in studying DDMS that function under conditions of indeterminacy as well as cooperative effects of center interaction in multi-level systems.

References

Aizerman M.A., Aleskerov F.T. *Choosing variants: fundamentals of theory.* M., "Nauka", 1990.

Arslanov M.Z. Scalarization of the task for constructing a set of optimal according to Slater solutions // *Automatics and Telemechanics.* 1997, No 8.

Berezovskiy B.A., Barishnikov R.M., Borzenko V.I., Kempner L.M. *Multicriteria optimization: mathematical aspects.* M., "Nauka".

Burkov V.N., Gorguidze I.I., Novikov D.A., Jusupov B.S. *Models and mechanisms for distributing costs and profits in a market economy.* M., 1997.

Burkov V.N., Danev B., Yenaleyev A.K. et al. Large systems: organizational mechanisms modelling. M., *"Nauka"*, 1989.

Burkov V.N., Zalozhnev A.U., Leontiev S.V., Novikov D.A., Chernishov R.A. *Mechanisms for financing regional development programs.* M., 2002.

Burkov V.N., Zalozhnev A.U., Novikov D.A. Graph theory in controlling organizational systems. M., *"Sinteg"*, 2001.

Burkov V.N., Yenaleyev A.K., Novikov D.A. Mechanisms for the functioning of socio-economic systems with reporting information // *Automatics and Telemechanics.* 1996. No 3. P.3-25.

Burkov V.N., Kondratiev V.V. Mechanisms for organizational systems functioning. M., *"Nauka"*, 1981.

Burkov V.N., Novikov D.A. How to manage organizations. M., *"Sinteg"*, 2004.

Burkov V.N., Novikov D.A. How to manage projects. M., *"Sinteg"*, 1997.

Burkov V.N., Novikov D.A.Active systems theory: condition and perspectives. M., *"Sinteg"*, 1999.

Guermeyer U.B., Yereshko F.I. Side payments in games with fixed sequence of moves // *"Zh.VM and MF".* 1974, No 14, P.1437-1450.

Guermeyer U.B. Games with non-contrary interests. M., *"Nauka"*, 1976.

Guilev S.E., Leontiev S.V., Novikov D.A. Distributed systems for decision-making in regional development management. M.: *"IPU RAN"*, 2002.

Gorelik V.A., Kononenko A.F. Theoretical game models for decision-making in ecological-economic systems. M., *"Radio and Communication"*, 1982.

Gubko M.V., Karavayev A.P. Harmonization of interests in array management structures // *Automatics and Telemechanics.* 2001. No 10. P.132-146.

Gubko M.V., Novikov D.A. Game theory in organizational systems management. M., *"Sinteg"*, 2002.

Danilov V.I., Sotskov A.I. Group choice mechanisms. M., *"Nauka"*, 1991.

Irikov V.A., Treniov V.N. Distributed decision-making systems. M., *"Nauka"*, 1999.

Kini R.L., Raifa H. Multicriteria decision-making: preferences and substitutions. M, *"Radio and Communication"*, 1981.

Kochieva T.B., Novikov D.A. Generic stimulation systems. M., *"Apostrophe"*, 2000.

Leontiev S.V., Messarovich M., Mako D., Takakhara., Hierarchical multilevel systems theory. M., *"Mir"*, 1973

Leontiev S.V. Models and methods for regional development management. M., *"Fizmatlit"*, 2002

Messarovich M., Mako D., Takakhara I., Hierarchical multilevel sysrems theory. M., *"Mir"*, 1973.

Mirkin E. Group choice problem. M., *"Nauka"*, 1974.

Moulen E. Co-operative decision-making: axioms and models. M., *"Mir"*, 1991.

Novikov D.A., Petrakov S.N. A course in active systems theory. M., *"Sinteg"*, 1999.

Novikov D.A., Petrakov S.N., Fedchenko K.A. Decentralization of planning mechanisms in active systems. *Automatics and Telemechanics.* 2000. No 6. P.126-137.

Novikov D.A. Stimulation in organizational systems. M., *"Sinteg"*, 2003.

Novikov D.A. Stimulation in socio-economic systems (generic mathematical models). M., *"IPU RAN"*, 1998.

Novikov D.A. Mechanisms for multilevel organizational systems functioning. M., Fund *"Management problems"*, 1999.

Novikov D.A., Tsvetkov A.V. Stimulation mechanisms for multielement organizational systems. M., *"Apostrophe"*, 2000.

Novikov D.A., Tsvetkov A.V. Functioning mechanisms for organizational systems with distributed control. M., *"IPU RAN"*, 2001.

Podinovsky V.V., Noguin V.D. Pareto-optimal solutions to multicriteria tasks. M., "Nauka", 1982.

Tsvetkov A.V. Stimulation in project management. M., *"Apostrophe"*, 2001.

Fudenberg D., Tirole J. Game Theory. Cambridge: *MIT Press*, 1995.

Green J., Laffont J.J. Incentives in public decision-making. Studies in public economics. Vol.1. Amsterdam: *North-Holland Publishing Company*, 1979.

Groves T. Incentives in teams // *Econometrica.* 1973. Vol. 41. No 4. P.617-631.

Groves T., Loeb M. Incentives in a divisionalized firm // *Management science.* 1979. Vol. 25. No 3. P.221-226.

Groves T., Radner R. The allocation of resources in a team // *J. of Economic Theory.* 1972. Vol.4. No 2. P.415-441.

Mas-Colell A., Whinston M.D., Green J.R. *Microeconomic theory.* N.Y.: Oxford Univ. Press. 1995.

Myerson R.B. *Game theory: analysis of conflict.* London: Harvard Univ. Press, 1991.

Chapter 3

SYSTEMIC OPTIMIZATION FOR REGIONAL PROGRAMS DEVELOPMENT

3.1. INTRODUCTION

The theoretical results gained in the organizational systems control theory [8, 10, 20] are used in making applied models which in turn are practically used in synthesis or modification of control mechanisms for real social-economic systems. From the point of view of the scale, regions are the largest control objects. In elaborating and implementing regional development programs the following methods are used: methods for a complex estimation of regions, competitive mechanisms for selecting enterprises and including them in the regional development program, methods for optimizing programs according to their cost, mechanisms for distributing financial resources including consent mechanisms and expert mechanisms [1, 2, 4, 7, 8, 11, 15, 16]. This chapter deals with a number of models and methods for systemic optimization of regional development programs (it implies a number of interconnected actions aimed at achieving the region development goal which, as a rule, are estimated according to a number of criteria) [7].

3.2. TAX BENEFITS AND LAX CREDITING

This paragraph deals with a number of mechanisms for supporting projects for reforming and restructuring (RRP). This support may be provided by the regional authorities to separate enterprises, by associations of enterprises to their divisions, etc. Specifically, mechanisms for tax benefits and mechanisms for financing RRP which differ from each other on the principles of making decisions concerning granting discounts by the center; such may be committees for supporting regional enterprises, the leadership of the association of enterprises, etc. And it is substantiated that granting tax benefits may be considered as giving credits with delay of repayment. Considerable attention is paid to the procedures for harmonizing interests of the parties interested in implementing RRP including the framework of distributed decision-making systems. The results of analyzing and comparing a number of mechanisms for tax benefits are generalized for a case when certain parameters are unknown to the center and are reported to it by enterprises.

Let us consider the following RRP enterprise model. The enterprise is characterized by nonnegative scalar volume of output $y \geq 0$. Before the enterprise starts the RRP implementation its fixed costs (for some fixed period of time considered to be solitary) are equal to c_0, specific variable costs (variable costs for producing a product unit) are equal to α_0, cost of a product unit realization - λ_0, peak output - $y_0^{\max}$. The enumerated parameters make it possible to determine the break-even point $y_0^{\min}$ - minimal production volume, at which gross receipt exceeds total costs (in our book independent numbering of formulae according to the chapters is accepted):

$$y_0^{\min} = c_0 / (\lambda_0 - \alpha_0). \tag{1}$$

If $y_0^{\min} > y_0^{\max}$, then the technologies should be changed (it is necessary to change nomenclature, to re-equip the enterprise so that to decrease fixed costs to the value c, to decrease variable costs to the value α, to increase maximal volume of output to the value $y^{\max}$, to improve quality, and consequently, the realization price to the value λ, etc.), as no organizational changes will be able to ensure profitability of the enterprise (see Figure 3.1) [15].

If $y_0^{\min} \leq y_0^{\max}$, then the optimal production volume is such a production volume which will ensure the following value of gross profit in a time unit: $\pi_0 = (\lambda_0 - \alpha_0) y_0^{\max} - c_0$. If profit tax rate is equal to γ, then net profit in a time unit is equal to

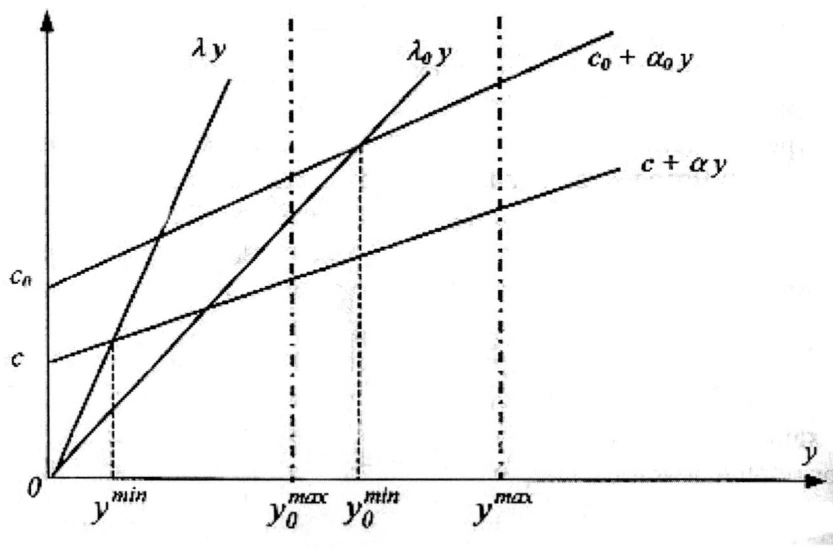

Figure 3.1. Break-even points.

$$\pi_{0\gamma} = (1-\gamma)\pi_0 = (1-\gamma)[(\lambda_0 - \alpha_0)y_0^{\max} - c_0].$$

Let us assume that the enterprise has non-budgetary debts (or budgetary debts subject to restructuring) G_0, then the period $T_{0\gamma}$ for going out of bankruptcy (i.e. period within which accumulated net profit will exceed debts) may be defined as

$$T_{0\gamma} = G_0 / \pi_{0\gamma}. \tag{2}$$

If the enterprise is granted tax benefits (i.e. it is exempt from part of taxes[3]), then the period T_0 of going out of bankruptcy is equal to

$$T_0 = G_0 / \pi_0 \le T_{0\gamma}. \tag{3}$$

Ex facte (see Figure 3.2.) granting tax benefits is equal[4] to deficiency of tax equal to γG_0 within the period of T_0.

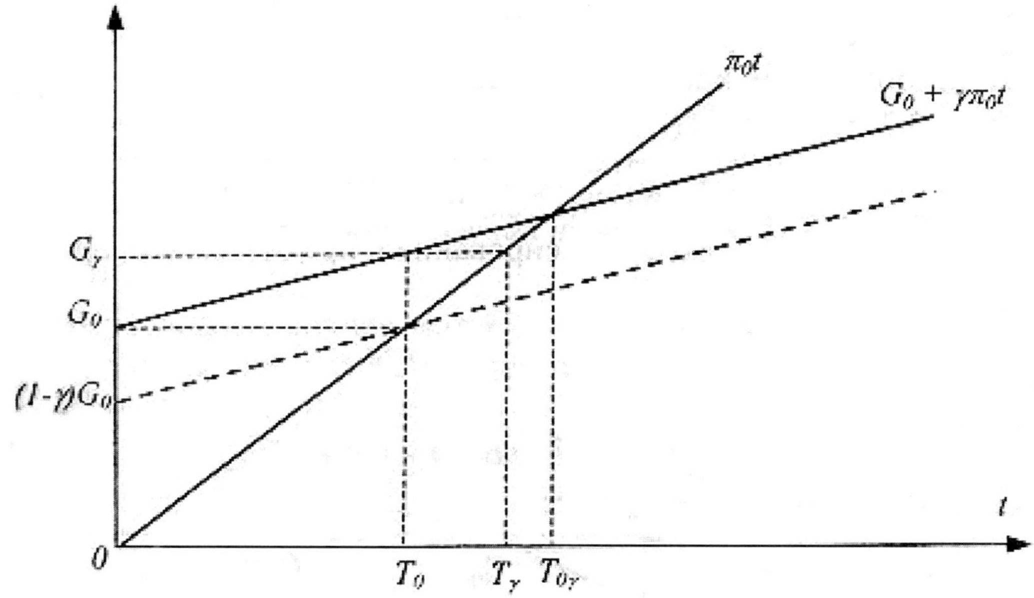

Figure 3.2. Dynamics of tax payments.

[3] For the sake of simplicity let us consider that there is only one tax – profit tax, and tax benefits correspond to a temporary complete exemption from discharge of this tax.

[4] As we consider dynamic models, it is necessary to state that all the financial values must be reduced to one point of time.

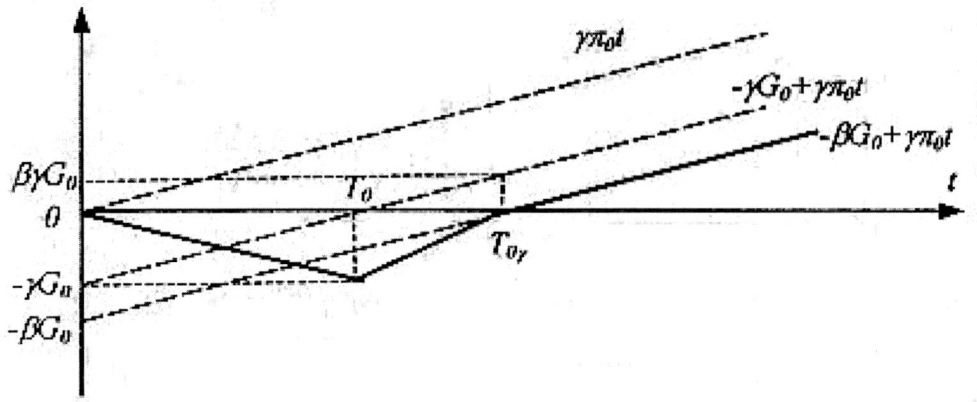

Figure 3.3. Dynamics of budgetary debts.

However, this conclusion is not quite correct from the standpoint of content, for it holds true under assumption that the enterprise pays taxes simultaneously with the offset of debt from net profit. In practice the situation is different – enterprises (having in a number of cases arrested accounts) frequently have to spend all their profit on offset of debt. As a result over time their debts grow by the size of unpaid taxes. Therefore, at the point of time T_0 (point of paying original debet G_0 without tax accounting) the enterprise accumulates "new" debts $G_\gamma - G_0 = \gamma G_0$ (see Figure 3.2), which will have to be paid at the point of time γT_0. But by the point of time $T_\gamma = T_0 + \gamma T_0$ a new debt is accumulated - $\gamma^2 G_0$, and so on. The point of time for paying all the debts (that are being accumulated) will be determined as a sum of geometric progression (see also expressions (2) and (3)):

$$T_{0\gamma} = T_0(1+\gamma+\gamma^2+...) = T_0/(1-\gamma) \tag{4}$$

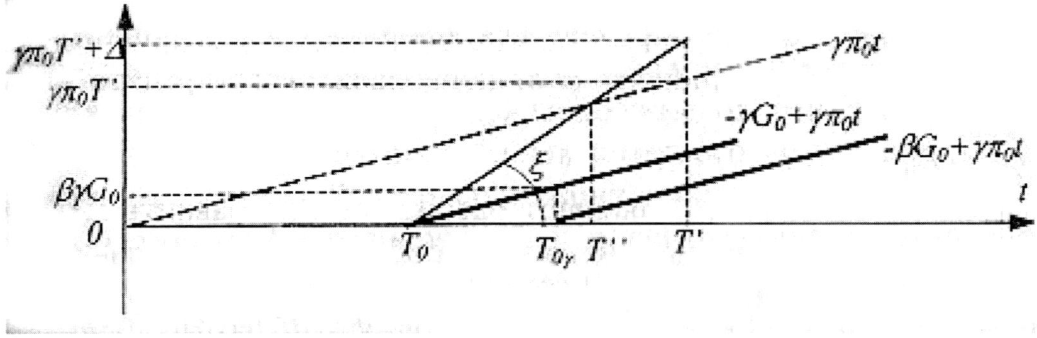

Figure 3.4. Dynamics of tax payments in case of tax benefits.

Consequently, during the period of time $T_{0\gamma}$ the enterprise is a debtor, and budgetary debts belong to the enterprise's most critical category of debts (see Figure 3.3, where $\beta = \gamma/(1-\gamma)(<1, \gamma < \frac{1}{2}))$

Therefore, offering preferences may be considered as an effective instrument for attracting investments. In such a situation it is advisable to use mechanisms for mixed financing and crediting [7].

Thus, in case of granting tax benefits the sum βG_0 may be considered as a benchmark for restructuring debts, i.e. giving tax benefits is equivalent to deficiency of taxes at the rate βG_0 within the period of time $T_{0\gamma}$. As it was mentioned above, this conclusion is not quite correct, for only high-position taxes should be considered (confer figures 3.3 and 3.4). The upper thick line in figure 3.4 corresponds to dynamics of budgetary payments in case the enterprise is exempted from all tax liabilities for the period of time T_0 discharge of non-budgetary liabilities. The bottom thick line corresponds to lack of tax benefits, the thin line corresponds to payment of taxes in case the enterprise is granted tax benefits γG_0 for the period of time T_0 provided they are paid by the point of time T', probably in case of nonzero per cent rate.

In the latter case we may consider that for the enterprise delay of repayment of budgetary debts (their transfer from the period of time T_0 to T') is equal to Δ (see figure 3.4). The point of time T'' characterizes the repayment of budgetary debts by the enterprise:

$$T'' = \frac{\gamma\pi_0 T' + \Delta}{\gamma\pi_0 T_0 + \Delta} T_0 . \qquad (5)$$

3.3. Mechanisms for Tax Recessions

Thus, from the point of view of the above-considered model, tax benefits may be interpreted as rediting with deferred payments. For analyzing its efficiency instruments for financial engineering [17] may be used. Getting such a credit may not be very efficient from the standpoint of a profitable enterprise not burdened with debts, however, we should underscore once more that in the situation we have in Russia at present for numerous enterprises on the verge of bankruptcy (according to the criterion of debts as well) it is the only way out of crisis.

The parameters of preferences (the date of beginning repayment, the cost of delay Δ, crediting rate ξ, etc.) are connected between themselves as follows (see also figure 3.4):

$$\xi = \frac{\gamma\pi_0 T' + \Delta}{T' - T_0} . \qquad (6)$$

The value Δ may be defined, for example, as additional fixed per cent - Ψ of the original credit γG_0, i.e.:

$$\Delta = (1+\Psi)\gamma G_0 . \tag{7}$$

The expressions (1)-(7) allow to consider and compare a number of mechanisms for tax benefits which differ from each other on decision-making principles concerning offering benefits to centers, whose role may be played by a committee for supporting regional entrepreneurship, a governing body of enterprises, etc. [15]. The center's decision-making principle is defined according to the efficiency criterion which the latter one seeks to maximize. Consequently, different efficiency criteria give birth to different mechanisms.

Let us consider an active system [10] (AS) consisting of n enterprises - active elements (AE) and one center. Let us assume that the center has an opportunity of granting R preferences, limited, for example, by the necessity to meet the regional budget usage of capacity requirement, creditability of the association of enterprises, etc. Thus, the next limitation must always have place (the bottom indexes designate the AE number):

$$\sum_{i\in Q} G_{0i} \le R/\gamma , \tag{8}$$

where $Q \subseteq I = \{1,2,...n\}$ is a subset of the set of all AE I comprising those ones that are offered tax benefits.

The center's task is to find set Q by which the efficiency criterion might be satisfied under balance (budget) limitation (8). Let us consider different criteria, i.e. different mechanisms for tax benefits.

Mechanism 1. Let us assume that the center is interested in removing a great number of AE from the category of budget debtors as soon as possible.

Technically this decision-making principle may be written down as:

$$Q = \{1,2,...,k_I\} , \tag{9}$$

where k_I, is a maximal number in ascending ordering T_i'' such as

$$\sum_{i=1}^{k} G_{0i} \le R/\gamma, \sum_{i=1}^{k+1} G_{0i} > R/\gamma . \tag{10}$$

The procedure of distributing benefits (see also competitive mechanisms in [1, 7]) consists in marshalling AE in ascending ordering of the dates of their removal from the category of budgetary debtors and granting them benefits starting with the first one until the resources at the center's disposal are exhausted.

Mechanism 2. Let us assume that the center is interested in maximizing its profits received as a result of benefits. As according to (7) this profit is proportional to the original debts, the competitive procedure of distributing benefits consists in marshalling AE in descending ordering of their original debts and granting benefits to them starting with the first one until the resources at the center's disposal are exhausted.

Mechanism 3. This mechanism is a compromise between the first and second ones. If the center is concerned that the enterprises both get out of bankruptcy as soon as possible and increase their profits (gains) due to tax benefits, it can calculate the efficiency of giving benefits to i-AE, for example, as follows:

$$Э_i = \gamma G_{0i} / (T_i'' - T_{0i}), i \in I ., \qquad (11)$$

marshal AE in descending order of their efficiency and grant benefits to them starting with the first one until the resources at the center's disposal are exhausted.

The above-mentioned three mechanisms of tax benefits belong to the category of simple competitive mechanisms [7] and, consequently, possess all the advantages and disadvantages appropriate to them. The main advantage is the simplicity of their application and transparence of the content interpretation and "rules of play" both for the competition organizers and competitors. The main disadvantage is the possibility of such situations when the efficiency of a simple competition turns out to be extremely low. Qualitatively it is caused by the fact that the application of efficiency ordering does not make it possible to consider all the variety of feasible versions (variants). As it is shown in (7), simple competitive mechanisms are less efficient than the so-called direct competitions, to whose category mechanism 4, analyzed below, belongs.

Mechanism 4. Let l_i - be the effect (result) of implementing i-project. We may ascribe to the effect (result) the following indicators (characteristics): the date of repayment of budgetary debts (the criterion of mechanism 1), the center' profit (the criterion of mechanism 2), some relative social-economic indicators (the relation of profit to the decrease of the period of time for repaying debts – see the criterion of mechanism 3), the number of the personnel of the enterprise who on getting out of debts receive a reward appropriate to the minimal basket of goods, etc. Let two-valued variable x_i, possess a unit value, if tax benefits are offered to i-enterprise, and it is equal to zero if benefits are not offered to it. Then the center's task is to define a set of projects which on being given benefits will maximize the effect (result)

$$L(x) = \sum_{i \in I} l_i x_i \qquad (12)$$

when the budget it limited (8), where $Q = \{i \in I \setminus x_l = 1\}$.

Task (12). (8) is a task about a knapsack [5]. The following statement holds true for this task.

Statement 3.1. Mechanism 4 possesses no less efficiency than respective mechanism 1, mechanism 2, and mechanism 3.

The correctness of this statement follows from the fact that to each solution received in mechanisms 1-3 corresponds a certain set Q (a set of competition winners), and in (12) maximization is conducted in all such possible sets, and the limitation (8) is the same for all the four mechanisms.

Let us consider an elementary numerical illustration demonstrating properties of mechanisms 1–4.

Illustration 3.1. Let there be 5 enterprises – aspirants to tax benefits. Data about them are given in table 3.1, the tax rate is equal to $\gamma = 30\%$, and the period of time for the repayment of debts $T' = 4$. The center has resource $R = 200$. The comparative efficiency of the mechanisms is given in table 2 (the sum of differences [remainders] between T' and T_i'' is maximized in mechanism 1).

It is seen that using a direct competition gives gain from 0 to 16 per cent of criterion of efficiency value.

A fundamental advantage of the fourth mechanism is also that applying (8) method of dynamic programming [5] with the purpose of solving task (12) we may receive the dependence of the criterion value on the size of the center's resource. This dependence allows to set and solve problems of defining the optimal (according to certain criteria) value of the summary (total) of tax benefits.

Table 3.1. Parameters of Enterprises

N	G_0	π_0	t_0	Δ	T''	x1	x2	x3
1	100	40	2.50	39	3.15	0	0	0
2	200	70	2.86	78	3.35	0	1	1
3	150	80	1.88	59	2.80	1	0	0
4	400	180	2.22	156	3.00	1	1	1
5	70	30	2.33	27	3.06	1	0	0

Table 3.2. Comparative efficiency of mechanisms 1-4

		Mechanism 4	Improvement (%)
Mechanism 1	3.15	3.64	16
Mechanism 2	234	253.5	8
Mechanism 3	276	276	0

Having solved the task of optimizing tax benefits according to some criteria, we received (see for illustration example 3.1) that different results follow from different criteria. To say it in other words, different mechanisms recommend offering tax benefits to different sets of enterprises (projects). There are no universal recepies – it is impossible to say a priori which of the mechanisms is "better". The way out is simultaneous use of a number of criteria followed by harmonization of decision-making in the framework of distributed system (see below and [13]).

While considering the mechanisms of tax benefits thus far we confined ourselves to a situation in which the center knew all the parameters of the enterprises. It is evident that in practice it is not always the case. Therefore, let us generalize the models under consideration for a case when certain parameters are unknown to the center and are reported to it by enterprises. Then a problem of data manipulation arises. Enterprises may report unreliable information trying to influence the decisions made by the center pro domo sua.

Let us consider how mechanisms 1–4 may be manipulated. Let us assume that such parameters of the model as: tax rate γ, original debts $\{G_{0i}\}$ are known beyond all doubt both to the center and AE. The size of specific profit π_{0i} is frequently not thoroughly observed by the center, therefore, let us assume that just estimates $\{s_i\}$ of these quantities are reported to the center by enterprises when the latter ones apply for participation in competitions for receiving tax benefits.

Let us consider mechanism 1. From expressions (5) and (7) we receive that the date of leaving the category of budgetary debtors for i-enterprise is determined as follows:

$$T_i^{"} = \frac{T'+(1+\psi)G_{0i}/\pi_{0i}}{2+\psi}, i \in I. \tag{13}$$

It should be noted that expression (13) does not depend on tax rate. It is known [11, 25] that AE game (play) in a mechanism of the type of mechanism1 is characterized by auction equilibrium, i.e. by such a set of strategies when all the AE – competition winners report profit estimates resulting in similar estimates $T_i^{"}$, whose difference is indefinitely small downward from the first (in ordering according to estimates $T_i^{"}$,) AE that did not find itself among the winners, which is characterized by time T_l, calculated below.

Let us assume that enterprises are interested first of all in being given with tax benefits (as it was stated above, it is the only opportunity of getting out of subbankruptcy or bankruptcy condition) and then (in case they receive tax benefits) it is an opportunity to understate the profits reported to the center. The latter tendency may be caused by the fact that credit repayments do not depend on profits (see expression (7)), and repayment delay increases the period of time and the size of surplus funds of the enterprise.

It follows from (13) that those enterprises will become winners whose minimal values $T_{0i} = G_{0i}/\pi_{0i}$, i.e. the following result is correct.

Statement 3.2. In mechanism (1) AE equilibrium reports look as follows:

$$s_i^* = G_{0i}(1+\psi)/(T_l(2+\psi)-T'), i \in Q, s_j^* = \pi_{0j}, j \in I \setminus Q, \tag{14}$$

where: $Q = \{1,2,...,k_l\}$, while marshalling AE according increase ordering of relations G_{0i}/π_{0i} number k_l, is defined by the expression (10) $T_l = T_{k+l}^{"}$.

The statement 3.2 means the following. Firstly, those enterprises will win in the competition, which are actually characterized by the minimum time $T_i^{"}$, (i.e. those enterprises which within minimal period of time are able to independently (from their own gains) repay foreign liabilities. Secondly, the total of interest outpayments for being granted a tax delay will be the same as in case of being fully informed (see expressions (7), (10) and (14)). Thirdly, under (13) and (14), underrating gains the enterprise takes the risk of not being given tax benefits, i.e. mechanism 1 makes it unfavorable to fundamentally underrate profits. Though, winners still manipulate information: under (14), having the same original debts, the greater is the profit AE has, the lower is its number and the greater the extent to which it distorts the information it reports to the center. And the more homogeneous the enterprises are (in the sense of T_{0i}), the narrower is the range of feasible distortions. And vica versa, the more heterogeneous are enterprises, the greater are the opportunities of monopolists (in the sense of T_{0i}) to distort information.

The above-mentioned effect of manipulating information by the competition winners in the absence of fines takes place in all competitive mechanisms [7]. To oppose this tendency either fines should be imposed (which is actually possible as decisions on offering benefits are based on enterprises' information concerning their future (planned) profits, and actual profits in the majority of cases are verified on the basis of financial accountability of enterprises, consequently, the size of fine may be based on the difference between actual and claimed (declared) profits or competitive mechanisms should be used together with certain antimonopoly mechanisms.

It is known [1, 7] that an effective way of "struggle" against monopolists is using non-cost-based mechanisms. The main idea of such mechanisms consists in choosing such dependence of the decision-making results on the parameters of the model which would make it unfavorable for AE to overrate (or underrate subject to the task) certain parameters. In the model under consideration monopolism shows itself in the possibility of underrating profits at the expense of resources in the difference $T_l - T_{0i}$. Consequently, it is necessary to take additional measures which will make it less favorable for the competition winners.

One of such measures may be resort to flexible dependence of credit rate or its repayment dates on AE parameters. It should be remembered that according to expression (7) AE payments to the center for having been granted a delay did not depend on AE profit, but were determined only by the size of the original debt and, consequently, by the size of benefits under the fixed date for the repayment of debts. Therefore one of the versions of building a mechanism impelling to report correctly (or to minimally underrate) the planned profit is introducing the dependence of the date of repayment or rate on reported profit, i.e. $\psi = \psi(s)$ or $T' = T'(s)$. An alternative competition is a two-stage one when at the first stage winners are chosen according to mechanism 1, and at the second stage winners are chosen according to the criterion of maximal profit among the enterprises getting additional benefits. A two-stage competition allows at the second stage (in the framework of which either again a competition procedure, or a mechanism for distributing fixed benefits [5], or a competition may be organized [5, 20], etc.) to differentiate winners (who on the strength of (14) at the first stage will report equal (identical) claims $T_l^{"}$) and thus increase the efficiency. At the same time holding the second round requires additional resources, therefore, in each specific case the center should correlate its costs with the effect produced.

Let us shortly consider the second mechanism in which the center aspires to maximize the summarized (total) profit gained as a result of offered benefits. As according to [7], this profit is proportional to the original debt observed by the center and competition procedure for distributing benefits does not depend on profit, information is not manipulated in this mechanism. If we assume that the center does not know the enterprises' debts and they report them to it, then mechanism 2 may be analyzed by analogy with mechanism 1 with the only difference that AE will not seek to underrate the size of debt it reports to the center, for in that case both the size of benefits and (in case of fixed profit) duration of benefits will decrease. In that case enterprises vica versa are interested in reporting overrated debts.

In the third mechanism according to (5) and (11) the center marshals AE in increased ordering of the following values:

$$E_i = \frac{\gamma(2+\psi)G_{0i}}{T' - \dfrac{G_{0i}}{\pi_{0i}}}, i \in I. \tag{15}$$

The estimate (15) of i-enterprise efficiency, calculated by the center, increases monotonously by G_{0i} and decreases monotonously by π_{0i}. AE equilibrium messages in mechanism 3 are calculated by analogy with the way it is done in statement 3.1 for the first mechanism. The drawback of mechanism 3 (as compared with mechanism 1) is that the competition winners are not interested in increasing profits even for the sake of winning it.

Mechanism 4, whose efficiency according to statement 3.1 is not lower, than that of mechanisms 1-3, unfortunately, is also subject to manipulation. In [7] it is proved that the guaranteed efficiency of direct competitions (to whose category belongs mechanism 4 is equal to ½. Thus, for example, in the conditions of illustration 3.1, while applying mechanism 4 in the task of minimizing the total time (corresponding to mechanism 1) the first AE may report profit estimate equal to 30. Then the composition of winners will not change, but the efficiency will decrease from 3.64 to 3.16.

Unlike the "continuous" task (whose solution is given in statement 3.2), in a discrete task arising in applying mechanism 4, a set of equilibrium strategies consists of more than one point – one and the same set of winners may have a great number of combinations of their strategies differing in a general case from authentic ones and being Nash equilibrium (see analogies in [18]).

For opposing the distortion of information the center must use more complex mechanisms, specifically, to induce AE to make decisions about messages under ambiguous conditions of the parameters of tax benefits mechanism. Let us build such a decision-making procedure.

As it was stated above, when solving the task

$$\sum_{i \in I} x_i l_i \to \max_{\{x_i\}} \tag{16}$$

under restriction

$$\gamma \sum_{i \in I} x_i G_{0i} \le R, \tag{17}$$

by applying the dynamic programming method [5], we have a possibility to find dependence $L(R)$ of optimal value of efficiency criterion (16) of restriction R, appearing in inequality (17). The knowledge of this dependence allows the center to offer the following decision-making procedure to AE (mechanism 5 for tax benefits): report values $s = (s_1, s_2, ..., s_n)$ of parameters unknown to me, and I will substitute them in tasks (16)-(17) and based on them I will calculate dependence $L(R, s)$. Then I will choose optimal (the criterion of optimality must be unknown to AE) value R^* and will offer benefits to those AE, which will turn out to be winners (will enter the solution to tasks (16}- (17)) provided $R = R^*$.

Statement 3.3. If the hypothesis of benevolence is fulfilled and AE use guaranteeing (according to a set of mechanisms for determining winners by the center) strategies, then mechanism 5 cannot be manipulated.

Proof of statement 3.3. Let us designate by $Q(R, s) \subseteq I$ the set of competition winners under restriction *R* and messages *s*. It should be noted that in a general (generalized) case this set is not unambiguously determined since different sets of AE are able to ensure the same effect value under equal costs. Therefore, let us consider that we are given a certain additional rule which unambiguously determines sets of winners in similar situations.

Let us fix a random number $i \in I$. Let it be that all the AE, except for i-AE, reported authentic information, i.e. situation of s_{-i}-play (game) is fixed for it. If s is fixed, dependence $L(R, s)$ is never decreasing piecewise constant right-continuous function *R*.

It follows from the solution properties of the task about a knapsack that under fixed environment for any *i*-AE s_i, message there are such values as R^- and R^+, and that $i \in Q(R^+, s_{-i}, s_i)$ and $i \notin Q(R^-, s_{-i}, s_i)$.

In other words under fixed environment for any AE message there are restrictions R under which it both finds itself among the winners and does not find itself among them. Consequently, on the strength of the hypothesis of benevolence i-AE is interested in reporting authentic information.

The result of the statement 3.3 cannot be considered as a promising one (in the sense of guarantee that AE reports authentic information), as it only ascertains that indefiniteness for AE is so great, that they will prefer to tell the truth for a mechanism can always be found in which AE will not find itself among the winners unless it reports authentic information.

Having discussed the issue concerning manipulation of mechanisms for tax benefits, let us more thoroughly discuss some procedures of coordinating criteria for choosing competition winners.

3.4. Distributed Systems for Making Decisions on Tax Benefits

As it is shown in the previous paragraph, mechanisms 1-3 apply different criteria for decision-making (center's behavior strategies) on offering tax benefits. It may be conventionally considered that mechanism 1 prescribes first of all to give benefits to high-profit enterprises deeply in debt, mechanism 2 prescribes to offer benefits to highly profitable enterprises deeply in debt regardless of their profitability (we mean - see above – that only such enterprises are considered which within the period of time T' are able to pay their debts to the center), mechanism 3 prescribes to give benefits to enterprises deeply in debt (it may take place for enterprises forming company towns, for objects of social sphere, etc., besides, marshalling enterprises according to value ordering reverse to (15), we will receive that highly profitable enterprises with slight debts will have priority, etc.).

Simultaneous application of several criteria characteristic of distributed decision-making systems (DDMS) is, on the one hand, very attractive, as allows the center to fully estimate the enterprise's activities. On the other hand, the application of several criteria causes certain difficulties and requires the elaboration of appropriate (valid) models and methods for decision-making.

One of the feasible mechanisms is connected with generating alternatives by means of solving multi-criteria task about a knapsack and analyzing these alternatives. Let us consider it more thoroughly.

Let us assume that there is a set of enterprises aspiring to get tax benefits. For this set a task family of the type (16)-(17) with different criteria is solved (the method of solving is thoroughly described below), then a final choice of winners is made based on the analysis of numerous estimates.

Let us consider the method of solving of the assigned task. Without loss of generality let us confine ourselves to the data of example 3.1.

Example 3.2. Let there be five enterprises whose data are given in table 3.1. Let us assume that the center simultaneously follows two criteria – the time t_1, when the enterprise gets out of the category of debtors (which is defined as the difference between $T'=4$ and T_i'') and profit Δ_i (or more exactly, income $\gamma G_{0i}(1+\psi)$, including planned debt payments), which the center gets providing this enterprise with tax benefits. As it was stated above, solutions to these tasks are different (irrespective of which mechanisms are applied: mechanism 1 and mechanism 2, or mechanisms 4 together with different effect indicators).

Let us apply the method of dynamic programming according to the following algorithm.

Let us build a network given in figure 3.5: let us from point (0;0) strike off two arcs which correspond to putting the first enterprise on the list of winners or not. The horizontal curve has a zero weight, the ordinate of the sloping (slanting) arc corresponds to the center's "costs" γG_{0i}. Let us put two numbers to correspond to the sloping (slanting) arc – additive criteria values

$[T_I - T_1''; \gamma G_{0i}(1+\psi)]$, given in the figure 3.5 beside the arc in square brackets through semicolon. Then, continuing the analogy for the second, third, fourth and fifth enterprises, we will receive in a general case $2^5 = 32$ variants.

If in a certain point two paths cross (in the example under consideration there was no such situation), then there are two sets of projects characterized by similar costs (which, as a rule, makes the dynamic programming method more efficient, than a simple complete enumeration (exhaustive search), and if one set Pareto-dominates over the other one according to criteria estimate, then the dominant estimates should be kept, and if there is no dominance, then both combinations of estimates should be considered in what follows (adding new projects).

We calculate costs (the first row of the right column in figure 3.5) for each of the final variants and, besides, applying the shortest route approach [5], we calculate criteria values (the second row, comprising couples of numbers in square brackets).

The advantage of the above-described method is that on adding new aspirants for getting tax benefits or expelling some of the existing ones there is no necessity to count over all the variants.

As a result in the example under consideration we receive 32 variants for choosing winners, each of them is described by two numbers – values of two criteria - K_1 and K_3, and costs minimum necessary for reaching (achieving) these values.

Then emerges a task of multi-criterion optimization (making decisions under many criteria), for whose solution there are a great number of thoroughly elaborated methods [14, 23, 24]. Let us consider a number of methods reflecting the specific character of the task we are solving.

The number of variants is quickly increasing with the increase of the number of aspirants[1][5](as two in the *n* power , where *n* means the number of aspirants). It is clear that even when there are not a great number of aspirants, the analysis of the content of all the variants is difficult, especially in case numerous criteria. Therefore it is necessary to elaborate procedures for cutting the number (preselection) of the variants to be analyzed. One of such procedures is the procedure used in the above-mentioned algorithm for sieving inefficient variants in the process of building a network, corresponding to the method of dynamic programming.

It should be noted that moving bottom-up (see the right column of data given in figure 3.5) the costs monotonously increase, but the criteria values in a general case on the strength of discreteness of the task change non-monotonously (monotony of estimates according to the criterion K_2 is caused by the fact that in the example under consideration it represents a linear transformation of costs). When there is only one criterion, the variants (alternatives of a great number of competition winners) at which monotony is disturbed, are not considered [5].

In case there are several criteria it is impossible to act so straightforwardly. The variants, which are Pareto-dominated by other variants (let us call this rule “rule No 1”) according to criteria K_1, K_2 and respective costs, should be excluded. By applying rule No 1 we can reduce the number of variants. However, this reduction does not always take place. In the example under consideration all the 32 variants are optimal according to Pareto. Therefore, another rule (let us call it “rule No 2”) states that only those variants should be kept, for which the increase of costs causes simultaneous increase of estimates according to all the criteria.

[5] It should be noted that the complexity of the procedure for generating variants does not actually depend on the number of criteria according to which alternatives are calculated.

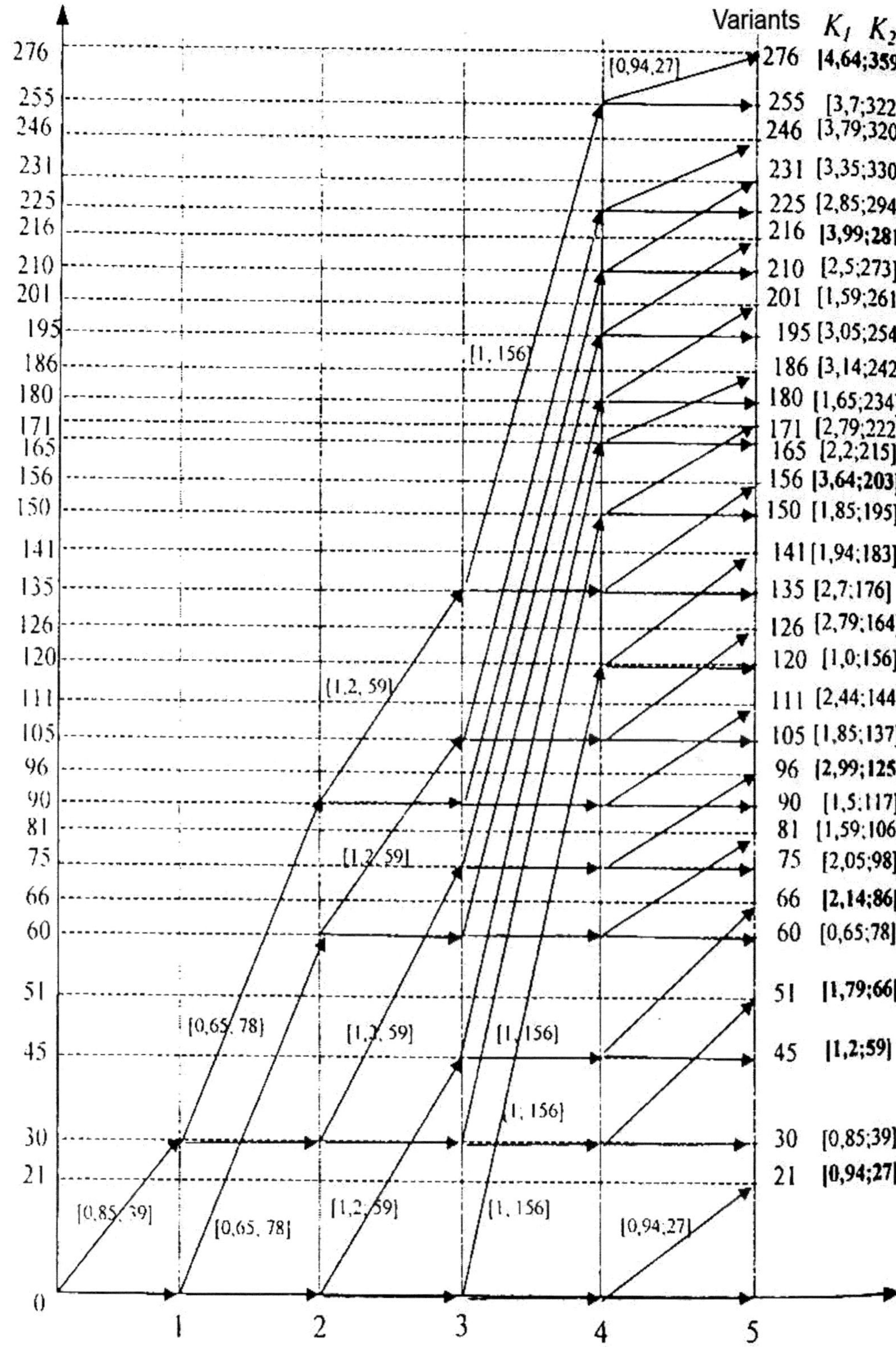

Figure 3.5. Method of dynamic programming.

For this purpose in case there are two criteria it is convenient to use the following method: let us put down points on the plane ($K_1;K_2$), which correspond to the selected variants and beside each point put down corresponding costs (diagram, reflecting the data of table 3.3, is given in figure 3.6). Examples of such an approach (in case when criteria are represented by effect and risk) are the so-called Risk-Efficiency-Costs (REC) diagrams [4].

Combined application of rules No 1 and No 2 as well as some other procedures are considered in example 3.3 below.

The application of rule No 2 in the example under consideration allows to reduce the number of variants we analyze down to seven marked out in extra bold in figure 3.5. Estimates of costs and criteria values concerning these variants are given in table 3.3.

Having reduced the number of variants, we can apply certain procedures of selecting the final set of competition winners for providing them with tax benefits.

The diagram we receive may serve as a base for discussing and coordinating the final variants (a great number of coordination procedures are described in [4, 14, 24]; see below as well) including situations when the center represents a distributed system, whose participants have different interests (different criteria priorities, etc.) [13, 15].

The information, given in figures 3.5-3.6, enables us to raise and solve a number of important problems (see the discussion and results in [4, 5]): to determine the minimum level of costs ensuring the specified estimate values according to certain criteria, to make decisions concerning expediency of getting a credit for financing competition winners, etc.

Having considered the problem of coordinating decisions in distributed systems and methods for solving these problems for lax crediting mechanisms, let us recall that we considered a set of enterprises that have debts and are able over time to independently repay their debts and pay to the center for the tax benefits they were given without essentially changing their structure, technology, etc. However, for the Russian reality along with the situation we have analyzed another one is also typical when enterprises in addition to big budget and off-budget debts are not profitable, i.e. unable to pay their debts, and over time it causes the increase of their debts. In that case it is necessary to implement projects for reforming and restructuring (RRP) which would make the enterprise profitable – see the model above. Therefore, now let us consider RRP models in the context of tasks for analyzing and synthesizing mechanisms for tax benefits.

Table 3.3. Variants for providing benefits

Costs	K1	K2
21	0.94	27.00
45	1.20	59.00
51	1.79	66.00
66	2.14	86.00
96	2.99	125.00
156	3.64	203.00
216	3.99	281.00
276	4.64	359.00

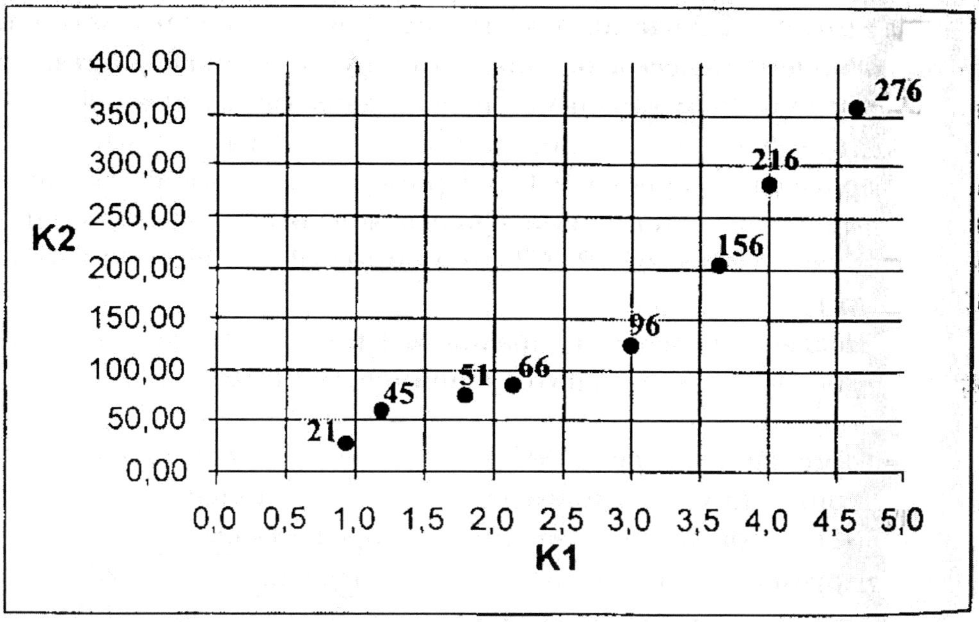

Figure 3.6. Variants of granting benefits.

3.5. PROJECTS FOR REFORMING AND RESTRUCTURING

The implementation of a project for reforming and restructuring the enterprise [15] requires certain costs (which may be considered like a certain debt) and some time later results in a certain effect. Therefore, all the results received above for tax benefits mechanisms without PRR, may be easily applied to those cases when benefits are given for carrying out RRP. For the sake of widening the variety of variants we will consider another model (connected with the first one), for which we will analyze mechanisms for making decisions concerning support for reforming and restructuring.

PRR Model

Let us specify the above-described model of the project for reforming and restructuring (PRR) of one enterprise which may consist in transforming technology (the change of nomenclature, technical re-equipment which makes it possible to decrease constant costs down to c, variable costs down to α, increase the maximum production volume up to $y^{\max}$, improve quality and, consequently, the selling price up to λ, etc.) – see figure 3.1. Let us assume that the implementation of PRR requires G costs (of original investments).

If after PRR is implemented it turns out that $y^{\min} \le y^{\max}$ is fulfilled, then the optimal volume of production will be that one which ensures the following value of gross profit in a unit of time: $\pi = (\lambda - \alpha)y^{\max} - c$. If the profit tax rate equals to γ, then net profit in a time unit equals to $\pi_\gamma = (1-\gamma)\pi = (1-\gamma)[(\lambda-\alpha)y^{\max} - c]$.

Let us confirm the assumption that the enterprise has a debt G_0, then the period of time T_γ for getting out of bankruptcy (i.e. the period during which net profit will exceed the sum of debts and the costs of PRR) may be determined as (see figure 3.7, in which the dynamics of financial indicators in the absence of PRR is expressed by an intermittent line and in the presence of PRR – by firm lines):

$$T_\gamma = (G + G_0)/\pi_\gamma . \tag{18}$$

The expediency of implementing RRP, from the standpoint of both AE and the Center, may be estimated according to the decrease of the period of time the enterprise needs for going out of prebankruptcy or bankruptcy, i.e. according to the decrease of the period of time for repaying debts. Therefore, we will consider conditions $T_\gamma \le T_{0\gamma}$ as RRP efficiency criterion. Substituting (2) and (18) we receive:

$$\pi / \pi_0 \ge 1 + G / G_0 . \tag{19}$$

Nonrecurrent PRR costs of G may be considered as a credit repayable within T_γ period of time. For credit see dash-and-dot line in figure 3.7 whose slope $\gamma\pi'$ is determined as follows:

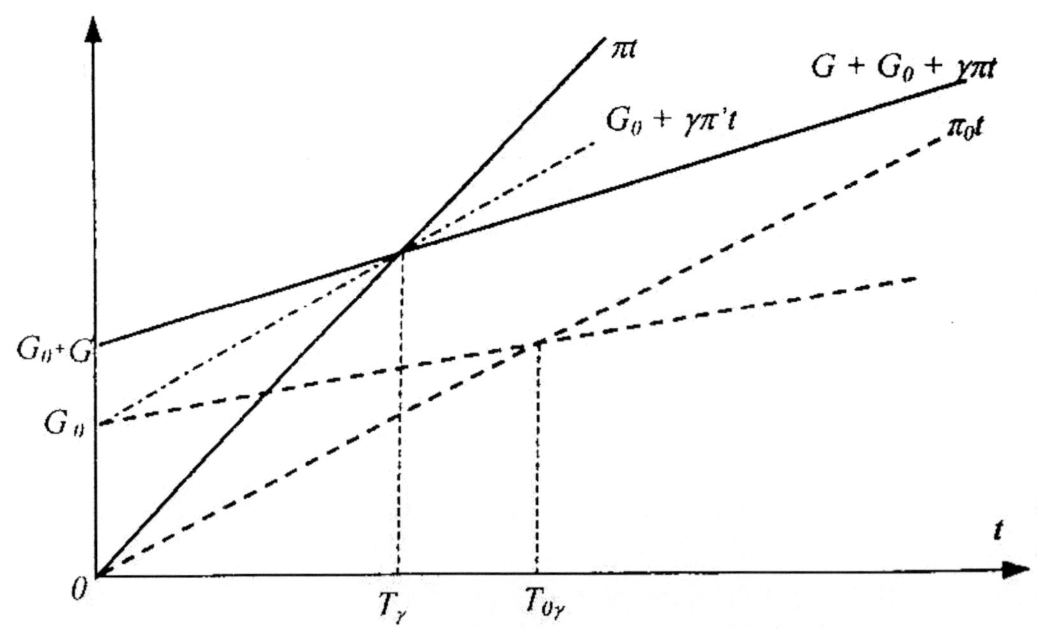

Figure 3.7. Dynamics of financial indicators after RRP

$$\pi' = \pi + G / T_\gamma, \tag{20}$$

therefore, value $\xi = G / T_\gamma$ may be interpreted as an equivalent credit rate (profit share used for repaying the credit):

$$\xi = \frac{(1-\gamma)G\pi}{\gamma(G+G_0)_\gamma}, \tag{21}$$

which monotonously increases proportionally to the increase of PRR profit and costs. Likewise, the efficiency E determined by the center as the ratio of the increase of tax payments to PRR costs, can be calculated: $E = \gamma(\pi - \pi_0) / G$.

Thus, PRR is characterized by G costs and the increase of payments to the budget $\gamma(\pi - \pi_0)$, and, in addition, by such derived quantities as: T_γ, ξ and E.

Having described PRR model for one enterprise, let us consider mechanisms for making decisions concerning the support the center offers to independent projects for reforming and restructuring carried out by a group of enterprises.

Mechanisms for Supporting RRP

Analogous to mechanisms 1-3 (simple competitions) in the above-mentioned model for making decisions about offering support to PRR is the following procedure: the center sorts AE in the order of decreasing or increasing of a certain criterion (for example, the time needed for repaying the debt, tax proceeds to the budget within a planned period, efficiency, etc.), and then gives credits to AE for implementing RRP, beginning with the first one until the resources are at the end. As was stated above, simple competitions may be considered as approximate (simplified) methods for getting an exact solution to the tasks about a knapsack corresponding to direct competitions. However, holding simultaneously several competitions (according to different criteria) is inexpedient and inefficient because of the necessity to break up funds (see the above statements and in [18] concerning the decomposition of optimization tasks in multilevel systems). It is easier to hold a multicriteria direct competition. Let us consider a practicable procedure for harmonizing interests reflected in different criteria (interest may be represented by different centers).

Let us assume that PRR are estimated according to two criteria: economic effect from their implementation (measured, for example, by the increase of payments to the budget – see above) and their ecological effect – their impact on the environment (measured, for instance, by the increase of pollution). It is clear that if criteria are monotonously connected, then the task is one-criterion and there is no need to harmonize interests. A problem arises when the improvement of value according to one criterion causes deterioration according another criterion.

The following procedure is offered for reducing the number of variants: first we select variants satisfying the existent restrictions, then nondominated ones are retained and, at last, the criteria (the center's interests) are harmonized which enables to retain one variant or a limited number of them (in the latter case a final decision must be made by the controlling organ superior to centers, or according to the earlier approved procedure. Let us more thoroughly describe the enumerated stages of the procedure for reducing the number of variants.

Let the following limitations be a priori given: R – for summary (total) costs and $\{R_i\}$ - for the minimum value of estimates according to criteria (for the sake of simplicity let us consider that centers' preferences are reflected in their tendency to improve the estimates according to all the criteria). Then a variant (a certain set of PRR) will be acceptable if it is characterized by summary (total) costs not exceeding R and estimates according to all the criteria not lower than those of corresponding minimum values $\{R_i\}$.

For generating a set of acceptable variants one can use a procedure for building a network analogous to the one used in the second example together with the branches and boundaries method. In this method adding or removing one RRP from a set of those being implemented corresponds to the branching of the tree of variants. And criteria for cutting off branches are as follows: costs exceeding maximum practicable ones or decrease of estimates according to at least one of the criteria to the minimum acceptable one. (It is interesting to note that if an empty set corresponds to the root of the tree, then there is every likelihood that first of all variants will be sifted (sieved) out because of low criteria estimates, and then because of the lack of resources. And if the implementation of all the PRR corresponds to the root of the tree, then there is all likelihood that first variants will be sifted out because of the lack of resources, and then because of low criteria estimates).

By applying rule №1 to the admissible variants, we will get a set of non-dominated variants. By applying rule 2 we will receive a succession of admissible non-dominated variants which is characterized by non-decrease of criterial estimates at the increase in expenditures.

Reducing a set of variants analyzed i.e. taking into account both limitations and Pareto-effectiveness in case this set consists of more than one variant, we must use additional procedures of multi-criterion choice, maybe by using harmonization of interests – see below.

Let us consider a numerical illustration.

Example 3.3. Let us assume that there are four projects; the data are given in Table 3.4.

In table 3.5 the variants of supporting PRR as well as values of expenditures and criteria are given («1» corresponds to the support of the given PRR in the variant in the line of 3.5 table; "o" – to lack of support). Estimates of variants (2, 6, 12, 16) obtained as a result of applying rule 2 (see above) are given in bold type. Dominated variants i.e. variants excluded according to rule 1 are italicized (2, 4, 7, 8, 13). It should be noted that the second variant emphasized by rule 2 is dominated i.e. it is ineffective.

Table 3.4. PRR Data in Example 3.3

Project number	1	2	3	4
Expenditures	10	20	15	25
Economic effect (K1)	4	5	6	7
Ecological effect (K2)	8	5	9	4

Table 3.5. PRR support variants in example 3.3

№	1	2	3	4	Expen-ditures	K1	K2	K
1	0	0	0	0	0	0	0	0
2	0	0	0	1	*25*	*7*	*4*	12
3	0	0	1	0	15	6	9	17.25
4	0	1	0	0	*20*	*5*	*5*	11.25
5	1	0	0	0	10	4	8	14
6	0	0	1	1	**40**	**13**	**13**	29.25
7	0	1	0	1	*45*	*12*	*9*	23.25
8	1	0	0	1	*35*	*11*	*12*	26
9	0	1	1	0	35	11	14	28.5
10	1	0	1	0	25	10	17	31.25
11	1	1	0	0	30	9	13	25.25
12	0	1	1	1	**60**	**18**	**18**	40.5
13	1	1	0	1	*55*	*16*	*17*	37.25
14	1	0	1	1	50	17	21	43.25
15	1	1	1	0	45	15	22	42.5
16	1	1	1	1	70	22	26	54.5

Variant estimates according to K_1 and K_2 criteria are given in Figure 3.8. The variants emphasized by rule 2 are marked by dots in bold; dominated variants are crossed out.

If the number of non-dominated variants is great (12 variants in this example), it is reasonable to introduce additional priorities of the criteria and calculate aggregated estimates.

For example, if the aggregated criterion $K=K_1+5/4K_2$ which reflects negligible priority of the second criterion over the first one is calculated , we will have only six non-dominated (from the point of view of expenditure and K criterion) variants presented in Figure 3.9.

If there are additional restrictions on expenditures and criterial estimates, then the set of admissible variants decreases: if R=50, $R_1=R_2=10$, then variants (6, 9, 10, 14) are admissible and non-dominated variant №14 being the best according to K criterion. If any other additional criteria are used at taking final decisions, the final option may change considerably. For example, if a variant characterized by maximum effect (in the sense of K criterion value) is to be chosen, than variant №10 will be selected.

Even in the model example it is shown that in a current discrete multi-criterion problem the final decision is extremely sensitive (i.e. unstable) to the choice of a priority system that determines the procedure of multi-objective choice. True, it is possible to use the whole "arsenal" of decision-making models and methods elaborated in multi-objective optimization from the theoretical viewpoint [13, 14, 23]. However, we should not forget that we deal with an active system in which quite concrete leaders (centers) stand behind the estimates of criteria; the leaders' (centers') preference should be reflected in the procedure of final choice of the variant for PRR set support. Therefore, let us consider procedures of harmonizing center interests.

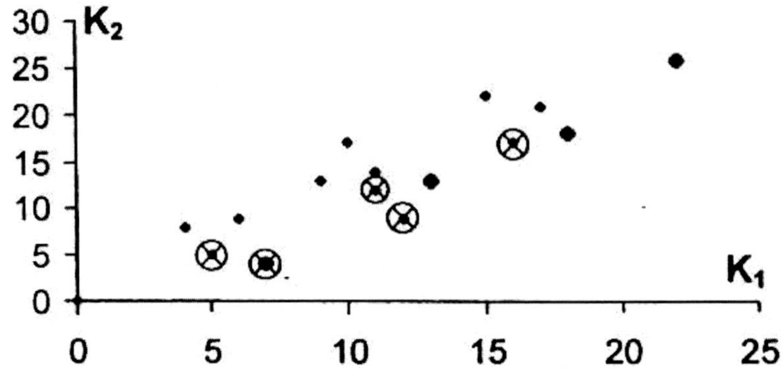

Figure 3.8. Estimates of variants in example 3.3.

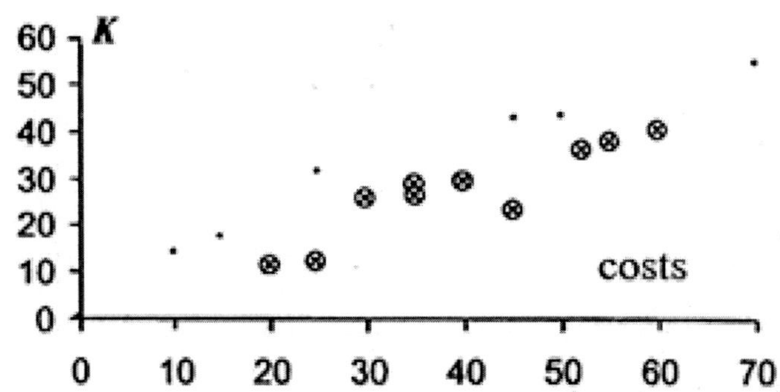

Figure 3.9. Estimates and cost inputs.

Procedures of Harmonizing Center Interests

As it was noted above, in distributed systems on PRR support, it is necessary to harmonize interests of those centers that maintain (i.e. that are interested in influencing the final decision and have the ability to do so) higher estimates according to definite criteria. Let us describe a possible procedure of harmonizing that is obtained as a result of solving the problem of motivational management [18, 22].

Let us consider a system consisting of n centers estimating m variants of PRR support. Let the utility of i center from realizing j variant be equal to $h_{ij}, i=\overline{1,n}, j=\overline{1,m}$. Let us fix 2 variants – j and k – and determine the "gain" of the i center from the "transition" of realizing j variant to k-variant:

$$\Delta_i(j,k) = h_{ij} - h_{ik}, i \in I = \{1,2,...,n\} \tag{22}$$

and the total gain of all the centers from this transition:

$$\Delta(j,k) = H_0(j) - H_0(k), \tag{23}$$

where $H_0(j)=\sum_{i=1}^{n} h_{ij}$.

Conceptually, $H_0(v)$ function may be interpreted as a utilitarian goal function of the "system" of n centers. $H_0(y)$ function is harmonized with the relation of domination according to Pareto in the following sense: if variant j Pareto-dominates variant k (according to the utilities of the centers and not according to criterial estimates), then $H_0(j) \geq H_0(k)$ (the opposite, generally speaking, is not true).

Let us introduce management into the model under consideration (the procedure of harmonizing center interests) i.e. let us add one control organ – the metacenter.

Introduction of a stimulation system $\{\sigma_{ij}\}$, corresponds to motivational management; taking into account this system the goal function of i center will be:

$$f_i(j)=h_{ij}-\sigma_{ij}, i \in I . \tag{24}$$

It appears that the interaction of the centers depends on $\sigma=\left\|\sigma_{ij}\right\|$ matrix. Let us assume that in the present task on motivation management there is a budget limit C on total stimulation.

First, let us study the harmonization of the interests of the centers without any budget limit $(C=+\infty)$. Let us fix two arbitrary variants j and k. In accordance with the results received in [20], the use of stimulation system by the metacenter

$$\sigma_i(\cdot)=\begin{cases}\Delta_i(j,k)-\delta_i, i=k \\ \sigma_i^N, \qquad i \neq k\end{cases}, \tag{25}$$

where "i=k" means the support of k variant by i center; $\sigma_i^N=\max\limits_j h_{ij}$ – the strategy of punishing the center for the rejection of k variant, $\delta_i>0$ – an indefinitely small strictly positive constant, stimulates all the centers to support k variant unanimously.

In expression (25) the first mode corresponds to the transfer of utilities, and the second one – to the punishment for individual deviations.

Let us proceed to the analysis of the balance (budget) limit. If utility transfers correspond to the inner stimulation that is closed in regard to a set of centers, then the sum of transfers must be non-positive to indefinitely small strictly positive constant $\delta=\sum_{i=1}^{n}\delta_i$. If the metacenter is able to obtain external funds or use its own funds at the rate C≥0, then the balance limitation (the so-called conditions of internal equilibration will be

$$\sum_{i=1}^{n}\sigma_i(j,k)=\Delta(j,k)=H_0(j)-H_0(k) \geq -C . \tag{26}$$

Thus, on the one hand, within the limits of the closed set of centers (at C=0), (26) is a condition for the non-negativity of transfer balance; on the other hand, this is, as it was mentioned above, the sufficient condition (taking into account (24)–(25)) for Pareto-dominance of variant *k* over variant *j*.

Let us analyze the role of budget limit. To achieve this let us fix an arbitrary variant k_0 and define the set of those variants that can be supported by centers (taking into account balanced motivational control by the metacenter) as an alternative to variant k_0:

$$P(k_0, C) = \{ j \setminus \Delta(k_0, j) \leq C \}. \tag{27}$$

It is clear that a set of *P(C)* variants that can be supported (as alternatives to any other variants) is

$$P(C) = \bigcap_{k_0} P(k_0, C) = \{ j \setminus H_0(j) \geq \max_i H_0(i) - C \}. \tag{28}$$

It is easy to demonstrate that when the metacenter uses stimulation system (24), any point of set (28) is Pareto-optimal.

Thus, the following result is correct.

Statement 3.4

At a given budget limit C any variant from set (28) can be realized by stimulation system (25).

Let us consider a question concerning the expediency of drawing external funds by the metacenter. Let the metacenter know for certain that the centers choose variant k_0 in the absence of control. Then $[\Delta(k_0, k) - C]$ is the indirect income of the metacenter from stimulating the centers to choose variant $k \in P(k_0, C)$. If *H(k)* is the metacenter's "own" income (or expenditures in case of a negative sign) earned from the realization of the corresponding variant, then the optimal value of the funds drawn may be found by solving the following optimization problem:

$$K(C) = \max_{i \in P(C, k_0)} [H(i) + \Delta(k_0, i) - \delta] - C \to \max_{C \geq)}. \tag{29}$$

Value

$$\gamma(C) = \max_{i \in P(C, k_0)} [H(i) + \Delta(k_0, i - \delta)] / C \tag{30}$$

can be considered as the ability of the system "to reinforce" (increase) the funds drawn; the augend stands for the investment of the metacenter, and the addend – for the investment of the centers ("tax" interpretations of the motivation management are given in [19]).

The procedure described makes it possible to determine the degree of the unmatched position of the interests of the centers and includes the method of linear convolution of criterial estimates (see example 3.3) as a special case. True, if the utility of each center is linear according to the corresponding criterial estimate, then $H(\cdot)$ is just the linear convolution of criterial estimates (within the limits of example 3.3 $h_{1j} = k_{1j}, h_{2j} = \frac{5}{4} k_{2j}$, are chosen where k_{ij} – is the estimate of *j*-variant according to *i* criterion). In a more general case, when the utilities of the centers are inseparable (each of them is, to a certain degree, interested in the increase of the estimates according to all the criteria), the described procedure also includes linear convolution as a special case.

3.6. SELF-FINANCE MECHANISMS

Under conditions of lack of circulating assets, which is characteristic of the current state of Russian economy, enterprises are not able to finance reforming and restructuring works on their own (each project on reforming – within the limits of the model considered – requires initial investments and a certain income can be brought in some time later). There is often no opportunity for getting borrowed current assets because in many cases the enterprises are involved in debts and credit cannot be secured. Therefore, the regional administration can finance reform projects or (which is more practicable at present and will therefore, be considered in the model) act as guarantors.

Let us consider the following model of the *active system* (AS) that consists of a controlling organ – the center – and *n* number of controlled subjects – *active elements* (AE). Each AE can carry out a certain measure (accomplish some work in terms of project control), characterized by cortege (c_i, d_i, τ_i) where c_i are expenditures necessary for starting *i* work, d_i – is an income after its completion, τ_i – the duration of work, $i \in I = \{1,2,..., n\}$ – a set of AE.

Let us assume that the operations are independent i.e. there is no technological interrelation between them which also determines the possible succession of their realization. As the income, drawn from the accomplished work can be used for financing new kinds of work, a problem of determining the optimal sequence of fulfilling this work arises. Financing mechanisms, in which the possibility of investing the obtained means in new operations, have been called *self-finance mechanisms* in [7]. In the afore-mentioned book a problem of determining the sequence of operations minimizing the maximum value of external funds attracted only once is considered. It was proved, that the solution for this problem (and at the same time, the solution for the problem of minimizing the total funds attracted) lies in the following sequence: first, profitable types of work should be carried out (i.e. those for which $d_i \geq c_i$) in the course of the increase in expenditures and then – unprofitable types (i.e. those for which $d_i < c_i$) in the course of the income decrease. These results can be directly used for solving problems in the model described in case the center itself finances the fulfillment of work. Therefore, let us consider in detail the case that has not been studied yet, when the center acts as a guarantor of paying the credit back by active elements and has the right to draw up the plan for carrying out work. Let us continue to work out the model in detail.

Let us denote the bank interest rate as α_0 (in a time unit) according to which it is possible to obtain borrowed current assets. To simplify, let us consider that credit security consists in it size.

Value $\alpha_i^0 = (d_i - c_i)/c_i$ characterizes the profitability of work i, and value $\alpha_i = (d_i - (1+\alpha_0\tau_i)c_i)/c_i = \alpha^0 - \alpha_0\tau_i$ *reduced profitability*[1] (reduced profitability can also be calculated in different ways [7]).

Let us consider the interest of the center in the following way. Let us assume that AE pay profit tax: $p_i = \{\beta\alpha_i c_i\}$, where β is the single rate of this tax. At the same time, in accordance with the suppositions that were introduced earlier, the center has to reserve means at the rate of $c_i, i \in I$ for the period of realizing *i* work.

If there are no limitations and the time of getting of tax payments is of no importance to the center (an idealized case), then it will be advisable to secure all the profitable (in the sense of the reduced profitability) operations, i.e. operations from set $Q_0 = \{i \in I | \alpha_1 > 0\}$; this would require the "freezing" of funds at the rate of $C^0 = \sum_{i\in Q_0} c_i$. However, there exist several criteria, that are taken into account by the centre in the process of decision-making. The priority of one or another criterion gives rise to the group of management problems that are considered below.

A certain schedule of guarantee commitments of the center as well as that of tax payments corresponds to every plan for work accomplishment; we will call these commitments reserves.

Let us consider, that all the expenditures and incomes have been reduced by the present moment i.e. to the moment of making decisions about the sequence of realizing the set of operations, which makes it possible not to consider discounting (this supposition takes place either at short-term projects or at taking into account of inflation in the crediting rate).

If t_i – is the moment of starting *i* work, then the financial balance of the center (in time) can be written down as follows (the difference from self-finance mechanisms studied in [5, 7] consists in the fact that the accumulation of expenditures does not take place):

$$f(t) = \beta\sum_{i\in Q_0}\alpha_i c_i I(t \geq t_i + \tau_i) - \sum_{i\in Q_0} c_i I(t \in [t_i; t_i + \tau_i)). t \geq 0 . \quad (1)$$

The time of work completion is determined by times $\{t_i\}$ as

$$T = \max_{i\in Q_0}\{t_i + \tau_i\} . \quad (2)$$

It is clear, that condition

$$\max_{i\in Q_0}\tau_i \leq T \leq \sum_{i\in Q_0}\tau_i , \quad (3)$$

is always fulfilled i.e. the time of project completion (the complex of operations) cannot be less than the duration of the most time–consuming operation (at the simultaneous and parallel fulfillment) and cannot be more that the total time for carrying out all of the operations (at successive fulfillment and without any breaks).

Maximum value of C_0 center *reserve* is defined by times $\{t_i\}$ as

$$C_0 = \min\{C \geq 0 \setminus \forall t \geq 0 f(t) \geq -C\}. \tag{4}$$

The dependence of the center *reserve* on the time is:

$$c(t) = \min\{0 : f(t)\}, \tag{5}$$

therefore, C_0 can also be defined as $C_0 = \max_{t\geq 0} c(t)$. The draft of possible diagrams of financial balance and center reserve is given in figure 3.10.

The aim of the center consists in carrying out profitable work with minimum reserves in the minimum period of time. However, the aim to minimize the reserve on the one hand and that to minimize the time, on the other hand, are at variance with each other. Therefore, for the purpose of revealing a set of rational variants (sequences of carrying our profitable operations) it is expedient to study possible combinations of time periods and reserves. Their extreme estimates can be received as a result of solving the following problems.

Problem 1. Find a set of sequences in carrying out operations (variants) on which the amount of centers reserves reaches its minimum, and out of this set choose the variant to which the minimum duration of the project fulfillment corresponds.

Problem 2. Find a set of sequences in carrying out operations (variants) on which the duration of the project fulfillment reaches its minimum and out of this choose the variant to which the minimum value of the centers' reserve corresponds.

Let us denote by (C_1, T_1) – the solution to problem 1 (values of maximum reserves and the project duration), and by (C_2, T_2) – the solution to problem 2. It is evident that $C_1 \leq C_2, T_1 \geq T_2$. The two estimates received are frontiers of the set of Pareto-effective variants (according to the criteria of maximum reserve and duration) – see figure 3.12, in which this set is given for example 3.4 considered below.

The process of solving each of the above-formulated problems can be divided into two stages. At the first stage a set of variants is determined on which the minimum of the value minimized in the first place is achieved. At the second stage the value, minimized in the second place, is optimized on a set of variants, received at the first stage.

The first stages of solving both problem 1 and problem 2 are realized simply enough.

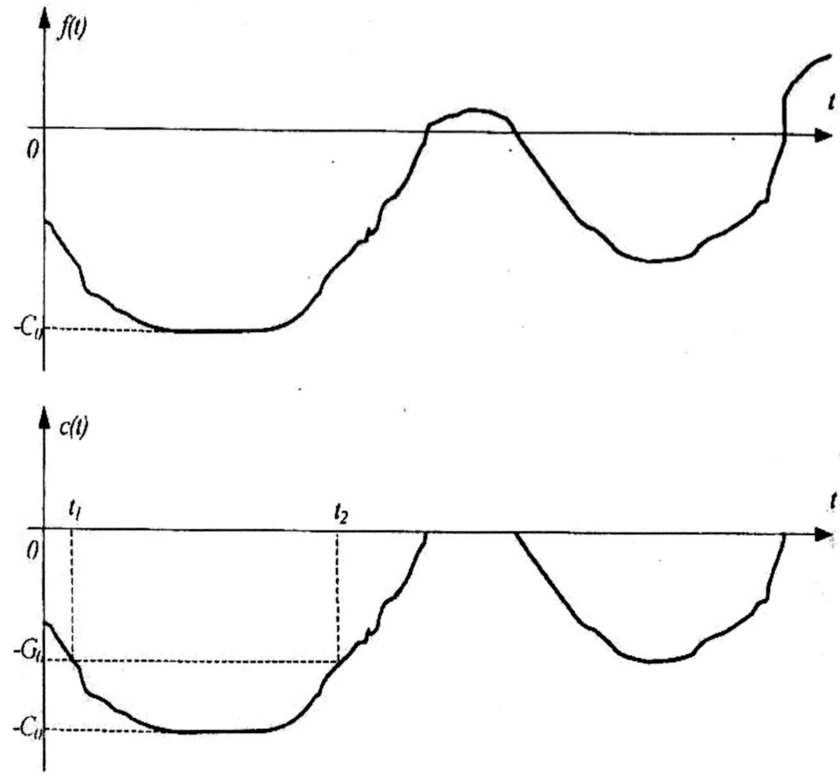

Figure 3.10. Example of the dynamics of financial balance and reserves of the center.

Let us consider a problem of search for sequential fulfillment of operations (variants) on which the center reserves' value reaches its maximum. This problem is a special case of the lecturer problem. It is clear, that in the optimal variants all works will be carried out in succession (i.e. neither of them should be carried out simultaneously) and without interruption. Consequently, an optimal order of carrying out work should be determined. To achieve this, let us make use of the results received in [5, 7]; according to these results carrying out profitable work in order of increase in their expenditures will correspond to the minimum of the maximum reserve (unprofitable operations i.e. operations from set $I \setminus Q_0$ will not be considered in future).

Let us put operations in order of the increase in expenditures: $c_1 \le c_2 \le ... c_m$ where $m = |Q_0|$. Putting

$$t_1 = 0, t_i = t_{i-1} + \tau_{i-1}, i = \overline{2,m} \tag{6}$$

and substituting the times of the work start (6) in (1) we get value C_I.

It should be noted that if the values of expenditures necessary for carrying out work are not known to the center and are reported to the it by AE, then AE are not interested in the distortion of expenditures, as they bear expenses themselves (they themselves cancel credits) and excessive expenditures will lead to such a situation that after the first stage of solving problem I the starting time of realizing the corresponding work will grow longer which is unprofitable for AE. To eliminate the profitability of manipulating with the information on expenditures on the part of AE the center should not, while informing AE on "the rules of the game", go into details concerning the rules of decision-making at the second stage.

The second stage of solving problem 1 (henceforth we will call it problem 1.2) consists in finding such starting times of work, that minimize the time needed for project completion on condition that the size of the maximum reserve does not exceed the known value C_i. This problem will be considered below.

Let us consider the problem of search for the execution sequence of operations (variants) at which the project accomplishment period reaches its minimum (the first stage of problem 2). It is clear that in the optimal variant all operations will be carried out simultaneously and without interruption. The project duration will be $T_2 = \max_{i \in Q_0}\{\tau_i\}$ i.e. it will be determined by the maximum duration of profitable operations. Hence, we only have to find an optimal order of work accomplishment (in future we will call this problem – problem 2.2), i.e. the maximum reserve of the center is minimized on condition that the project duration does not exceed the known T_2 value. This problem will be studied below.

Let us consider the following example illustrating the solution to the problem of optimizing the execution sequence of operations.

Example 3.4. Let there be 5 operations whose data are given in table 3.6, $\alpha_0 = 20\%, \beta = 40\%$. It is shown in the table that operation 5 is unprofitable according to the criterion of reduced profitability (though it is profitable if we do not take into account interest payments on credit) so we can exclude it from the subsequent consideration.

The solution of the first stage of problem 1 gives us the following sequence of operations fulfillment (without simultaneous accomplishment and interruptions): 1–2–3–4.

The maximum reserve is equal to $C_0=33,2$ and the project duration is $T_{10}=12$ time units (see Figure 3.2). At the second stage of solving problem 1 we should find the execution sequence of operations (operations in round brackets are fulfilled simultaneously): (2; 1)–(2; 3)–2–4, at this sequence the volume of maximum reserve will not change and the project duration will be reduced to $T_1=9$ time units (see the time in bold type in Figure 3.11 – the figures at the arcs correspond to operation numbers). After optimizing the time the diagram of reserves has become more "uniform".

The solution to the first step of problem 2 consists in the simultaneous beginning of accomplishing all the operations which requires maximum reserve $C_{20} = 80$ and time $T_2 = 5$ of the time units. At the second step of solving problem 2 it is possible to reduce the volume of the maximum reserve to $C_2 = 70$ without increasing the time needed for the project accomplishment.

Ploying the results of the solutions to problems 1 and 2 (points (C_1, T_1) and (C_2, T_2)) on plane $C_0 0T$ in Figure 3.12 we will get a crosshatched estimate (it is clear that in a discrete problem a finite set of variants is possible) of a set of possible combinations of maximum reserves and periods of time.

Table 3.6. Parameters of operations in example 3.4

№	c_i	d_I	τ_i	α_I^0	α_i
1	10	20	1	1.0	0.8
2	10	25	4	1.5	0.7
3	20	30	2	0.5	0.1
4	40	90	5	1.25	0.25
5	30	40	3	0.33	-0.27

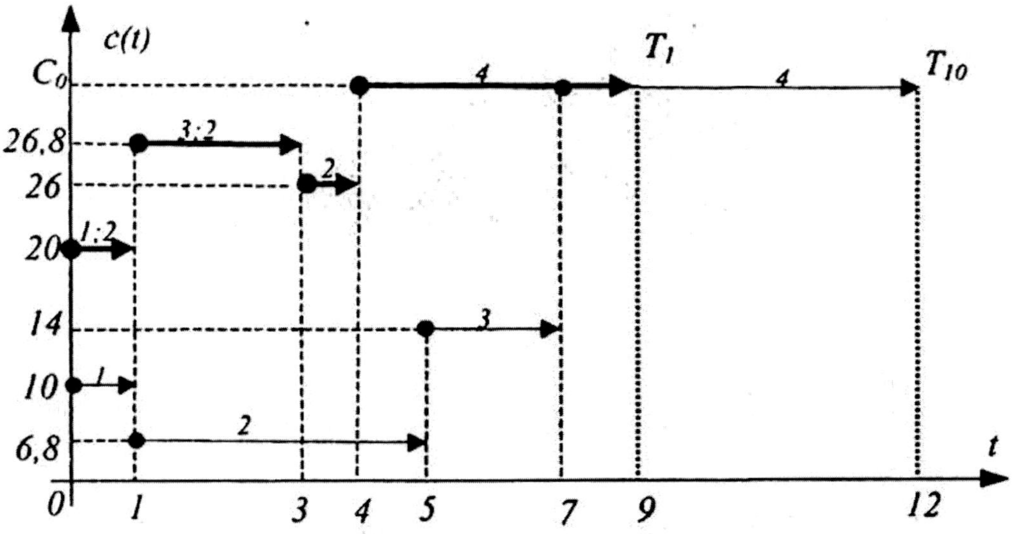

Figure 3.11. Optimal sequence of operations in problem 1.

Above we have described methods of solving problems concerning minimizing maximum reserves of the center (the time of project accomplishment) to which two points in Figure 3.12 correspond. The problem of enumerating all the variants having Pareto-effective estimates of duration and reserves is more general. In an even more general case the dynamics of the financial balance of the center can be estimated not only by the volume of maximum reserve but also by a certain functional reflecting the priorities of the center

Let us consider methods of solving a problem for minimizing the reserve on condition that the directive term T for realizing all operations (reform projects) has been determined.

Let n_K number of operations, fulfilled at every period $k = 1 \div T$ (it is clear that $1 \leq n_K \leq n - T$, where n – is a number of operations of a regional program) be determined. In this case the priority of operations rule remains valid; according to this rule the operations are set in the order of increasing c_i (the proof is analogous to the proof for the sequential chain of operations). In this way, the problem is reduced to determining the n_K number of operations fulfilled in period k. To solve this problem let us apply the method of dynamic programming.

Let us plot a grid on the plane. Let us plot k=1, T periods on the ordinate. Let us plot N_K numbers – the number of operations fulfilled for the first k periods is $k \leq N_K \leq N$.

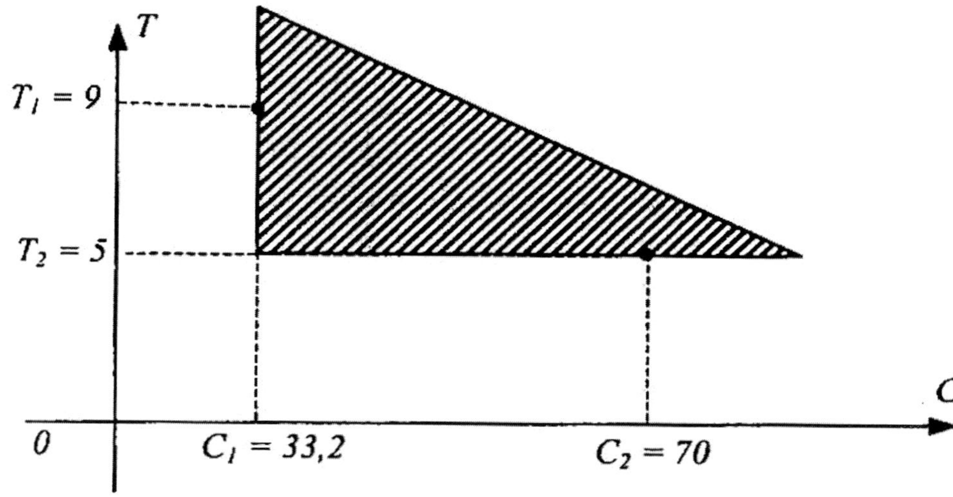

Figure 3.12. Set of permissible combinations of maximum reserves and time periods.

The vertexes of the network correspond to the points with coordinates $(k,\ N_K)$. From vertexes $(k\!-\!1,\ N_{K-1})$, $k=1\div T$ arcs proceed to all the vertexes $(k,\ N_K)$ so that $N_K > N_{K-1}$. It should be noted that each arc $(N_{K-1},\ N_K)$ is in correspondence with a quite definite set $Q\ (N_{K-1},\ N_K)$ of operations that are fulfilled in k period, namely, these are operations with numbers from $N_{K-1}+1$ to N_K inclusive.

Let us denote by

$$D(N_{k-1}) = \beta \sum_{q=1}^{N_{k-1}} \alpha_q c_q \,.$$

the means that had been received by the center by $(k\!-\!1)$ period inclusive as a result of fulfilling the first N_{K-1} works, and by

$$C(N_{k-1}, N_k) = \sum_{N_{k-1}+1}^{N_k} c_q$$

the means that the center must reserve to fulfil the operations with numbers from $(N_{K-1}+1)$ to N_K inclusive in k period. Let us designate the difference

$$L(N_{k-1}, N_k) = C(N_{k-1}, N_k) - D(N_{k-1})$$

as the length of the arc $(N_{K-1},\ N_K)$.

Let is note that any path in the net connecting the input to the output is in correspondence with a certain plan of project realization and vice versa.

The model constructed makes it possible to solve a number of problems concerning the optimization of the project realization plan. Thus, the problem of determining the minimum reserve under conditions of all the works being fulfilled during T time is reduced to determining the path that connects the initial vertex with the terminal one, the path the maximum arc length of which is minimum.

Example 3.5. Let there be 4 works the data on which are presented in Table 3.7.

Let us assume that $T=4$. The network of possible variants of carrying out operations is presented in Figure 3.13. It should be noted that we receive optimal plans for carrying out operations for any T. Thus, at $T=3$ in the optimal plan of the first period, the first operation is carried out, in the second – the second and the third ones, and in the third – the fourth one (the minimum volume of the reserve is equal to 10). At $T=2$ in the first period the first and the second operations are fulfilled, and in the second – the third and the fourth ones. The paths corresponding to the optimal plans are marked in bold type in Figure 3.13.

The constructed network has an interesting peculiarity. Arcs (N_{K-1}, N_K) and (N_{q-1}, N_q) for which $N_{K-1}=N_{q-1}$ and $N_K=N_q$, have similar lengths. This makes it possible to simplify the network. In figure 3.14 the set received from the set of figure 3.13 is given.

Table 3.7. Parameters of works in example 3.5

№	1	2	3	4
C_I	10	10	20	40
P_I	20	25	30	90

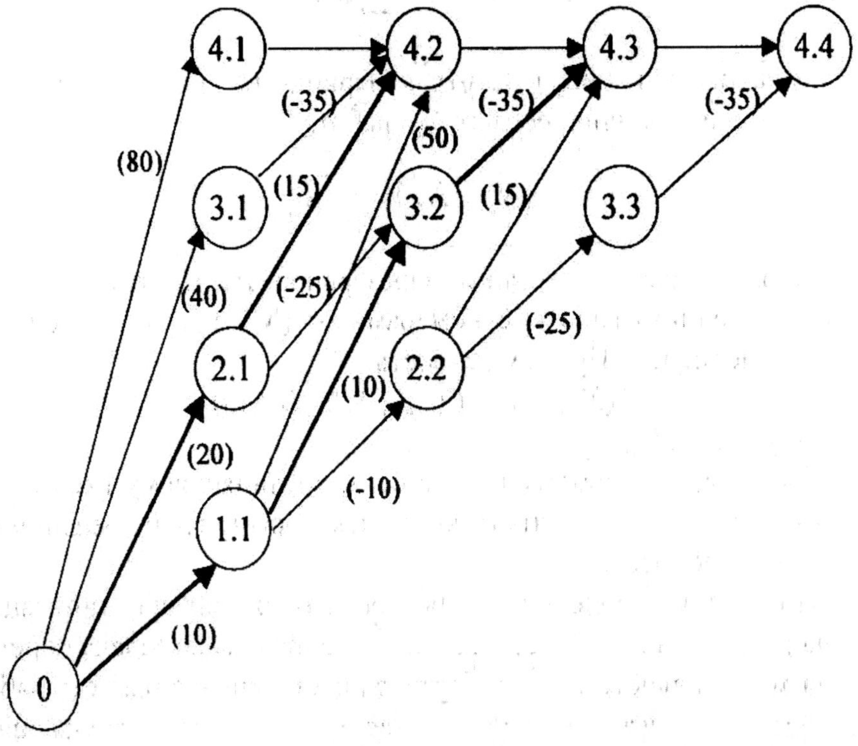

Figure 3.13. Network of variants of operations accomplishment.

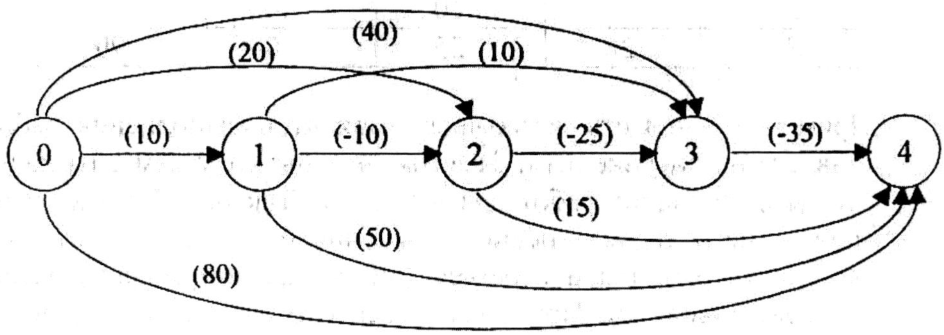

Figure 3.14.

Each i vertex of this network corresponds to the state of the project in which the first i operations are carried out. Every path of the net connecting the initial vertex with the terminal one, determines a certain plan of operations accomplishment and vice versa. The number of arcs of the path determines the project duration. If arc (i, j) belongs to path μ and is the k arc of the path, this means that operation from $(i+1)$ to j inclusive are fulfilled in k period. Thus, the problem is reduced to determining the path consisting of T arcs, the maximum arc length which is minimum.

The inverse problem is that of determining the path with the minimum number of arcs; in this path the lengths of all the arcs do not exceed the predetermined value (the reserve size). To solve this problem it is sufficient to remove all the arcs the lengths of which exceed the predetermined value and to determine the shortest (according to the number of arcs) path in the network received. Thus, the size of reserve being $S=10$ we get path $\mu(10)=(0,1,3,4)$, at $S=15$ – path $\mu(15)=(0,1,2,4)$ and at $S=20$ – path $\mu(20)=(0,2,4)$. It is evident that paths $\mu(10)$ and $\mu(20)$ are Pareto-optimal; for them the project duration constitutes 3 and 2 periods, respectively.

If the time periods for work are different, then the problem does not have such effective solution algorithms. However, on the basis of the algorithms described above, it is possible to receive estimates from below of either the volume of the reserve or the duration of the project and use them in the algorithms of branch types and limits. Let us illustrate the idea of this approach on example 3.5. Let $T=5$. In this case operation 4 should be carried out in all the periods. We have two variants for operation 2; it starts either in the first period or in the second one (it is supposed that operations are carried out without any interruption).

Let us consider the first variant. Let us divide operation 3 into two operations of single duration with expenditures in 10 each and 0 payments to the center for the first operation and 2 units for the second one. By solving the problem of minimizing the reserve at single duration of operation 1 and two sub-operations 3, we get that the minimum reserve value is equal to 66,8 units.

Let us consider the second variant, when the second operation starts in the second period. For this case, the minimum reserve size is also equal to 66,8. Therefore, there are several optimal solutions. One of them is as follows: in the first period operations 1 and 4 start, in the second – operations 2 and 3. It should be noted that the solution, received in example 3.1 is not optimal (the reserve volume is equal to 70).

Having considered self-finance mechanisms, let us discuss their basic advantages and disadvantages.

The considerable advantage of self-finance mechanisms consists in the fact that, firstly, the project is realized owing to external funds; secondly – owing to the maximum use of internal reserves. Indeed, as compared with the project accomplishment at the expense of the center, to the AE external funds (investments) are attracted .

In the example of the regional development program external investments whose guarantor is the administration of the region, on the one hand, make for the realization of the project on reforming and restructuring enterprises; on the other hand, guarantee reserves also stay in the region and what is more, they may and must be used efficiently (see below).

The main disadvantage of the self-finance mechanism that uses external funds is the need to "freeze" the reserve fund. Let us consider as to how tough the requirement concerning the complete freezing of the funds is i.e. what possibilities the center has concerning the *operational management of the reserve fund.*

The specific character of guarantee investments consists in the fact that the center undertakes to secure the repayment of credit, either completely or partially, if a certain AE is not able to do without assistance. The volume of the reserves $c(t)$ obtained as a result of solving the self-finance problem (extreme estimates are received following the solution of problems 1 and 2), characterizes the dependence on the time of maximum commitments on the part of the center i.e. at time t the center may need $c(t)$ funds at most. But in case of the successful fulfillment of operations i.e. at getting the planned income these means may be unnecessary. To show this uncertainty, let us introduce the concept of *risk* r_1 – the probability of not getting the planned income in i operation.

Let us assume that apart from the project under consideration, the center has a number of investment projects in which it can invest. Then the *reserve fund operational management* problem consists in the search of the most optimal distribution of the fund among investment projects (e.g. according to the maximum profit criterion or the minimum loss of profit or maximum of socio-economic effect for the region in case of the regional development project and so on) taking into account certain dynamics of reserves and risk factors. A lot of investigations are concerned with solving this problem [4,5,7,15]. The results obtained in these operations, in the first place, in operations on studying insurance mechanisms [6,7], can be used for solving the problem of the reserve fund operational management in the models of organizational projects as well as those of reforming and restructuring projects.

As the volume of the claimed reserves of the center is a random variable, creating risk management mechanisms is one of the aspects of the reserve fund operational management (insurance mechanisms are considered to be the subclass of risk management mechanisms [6]).

The problem of self-financing projects of regional development can modified in different directions. Thus, we supposed that nonrecurring income results immediately after work completion. In many cases (e.g. in case of new production output) after completing work the income is drawn along with the output and sale of the production. Thus, we supposed that expenditures were necessary from the moment of starting work. It is interesting to consider cases when diagrams of expenditures are set beforehand. Finally, as it was noted above, different optimality criteria are possible.

3.7. Conclusion

In this chapter the mechanisms of tax benefits and financing of PRR are described and studied, models and methods for harmonizing interests in distributed decision-making support systems are analyzed; properties of self-finance mechanisms are studied. Preferential taxation mechanisms, PRR crediting mechanisms and self-finance mechanisms are aimed at the increase in effectiveness (which is understood in the broad sense of the word as a complex criterion) of enterprises functioning, that finally leads to the increase in the effectiveness of the regional development program. These mechanisms can be used in succession, for instance first the center takes decisions to support PRR, then the self-finance mechanism is created; later, a problem of tax benefits is studied. However, such decomposition may lead to the drop in effectiveness, therefore, in a general case, the expediency of supporting certain PRRs with regard of self-finance possibility and granting benefits should be analyzed at the same time.

References

Andronnikova N.G., Barkalov S.A., Burkov V.N., Kotenko A.M. Models and methods for optimizing regional development programs. M., *"IPU RAN"*, 2001.

Andronnikova N.G., Burkov V.N., Leontiev S.V. Complex evaluation in regional management tasks. M., *"IPU RAN"*, 2002.

Burkov V.N., Danev B., Yenaleyev A.K. et al. Large systems: organizational systems modeling. M., *"Nauka"*, 1989.

Burkov V.N., Javakhadze G.S. Economical-mathematical models for managing branch-wise industry. M., *"IPU RAN"*, 1998.

Burkov V.N., Zalozhnev A.U., Novikov D.A. Graph theory in controlling organizational systems. M., *"Sinteg"*, 2001.

Burkov V.N., Zalozhnev A.U., Koulik O.S., Novikov D.A. Insurance mechanisms in socio-economic systems. M., *"IPU RAN"*, 2001.

Burkov V.N., Zalozhnev A.U., Leontiev S.V., Novikov D.A., Chernishov R.A. *Mechanisms for financing regional development programs.* M., 2002.

Burkov V.N., Novikov D.A. How to manage organizations. M., *"Sinteg"*, 2004.

Burkov V.N., Novikov D.A. How to manage projects. M., *"Sinteg"*, 1997.

Burkov V.N., Novikov D.A.Active systems theory: condition and perspectives. M., *"Sinteg"*, 1999.

Guilev S.E., Novikov D.A., Leontiev S.V. Distributed systems for decision-making in regional development management. M.: *"IPU RAN"*, 2002.

Gubko M.V., Novikov D.A. Game theory in organizational systems management. M., *"Sinteg"*, 2002.

Irikov V.A., Treniov V.N. Distributed decision-making systems. M., *"Nauka"*, 1999.

Kini R.L., Raifa H. Multicriteria decision-making: preferences and substitutions. M., *"Radio and Communication"*, 1981.

Leontiev S.V., Masiutin S.A., Treniov V.N. Success strategy: generalization of Russian plant facilities reforming experience. M., "OOO" Printing House *"Novosti"*, 2000.

Leontiev S.V. Models and methods for regional development management. M., *"Fizmatlit"*, 2002.

Marshall D.F., Bansal V.K. Financial engineering. M., *"Infra-M"*, 1998.

Novikov D.A. Mechanisms for multilevel organizational systems functioning. M., Fund *"Management problems"*, 1999.

Novikov D.A., Petrakov S.N., Fedchenko K.A. Stimulation in project management as a backbone factor. Proceedings of the international symposium *"Sovnet-99"*, M., 8-11 September, 1999.

Novikov D.A. Stimulation in organizational systems. V., *"Sinteg"*, 2003.

Novikov D.A., Tsvetkov A.V. Stimulation mechanisms for multielement organizational systems. M., *"Apostrophe"*, 2000.

Novikov D.A., Tsvetkov A.V. Functioning mechanisms for organizational systems with distributed control. M., *"IPU RAN"*, 2001

Podinovsky V.V., Noguin V.D. Pareto-optimal solutions to multicriteria tasks. M., *"Nauka"*, 1982.

Trakhtenguerts E.A. Computer-supported decision-making. M., *"Sinteg"*, 1998.

Myerson R.B. *Game theory: analysis of conflict*. London: Harvard Univ.Press. 1991.

Conclusion

Systemic optimization is a comparatively new concept and different authors interpret it in different ways. This book reflects its authors' point of view, especially, concerning the role of systemic regularities in solving problems of systemic optimization.

The authors hope that discussions concerning both the concept itself and the approaches and methods applied will promote the development of this topical field of organizational systems control.

About the Authors

Iveri V. Pranguishvili: director of V.A.Trapeznikov Institute for Management Problems at the Russian Academy of Sciences (Moscow), Honored Science Worker of the Russian Federation, doctor of technical sciences, professor, academician of the Georgian Academy of Sciences , academician of a number of international and foreign academies.

I.V. Pranguishvili is an outstanding specialist in the sphere of control systems, information and computer science, author of more than 350 scientific works including 15 books, monographs, one discovery and 30 inventions.

The main directions of his scientific activities are as follows: working out the theory and principles of systemic approach and general systemic regularities according to which different complex systems function; elaborating new (background) principles for revealing and recognizing moving objects of different nature; developing a theory and principles for constructing distributed systems of large-scale objects control as well as creating multiprocess control computer systems with a parallel and reconfigurable structure.

The results of I.V. Pranguishvili's research have been successfully implemented in different spheres for creating complex automated objects controlling large-scale systems. More than 30 doctor's and candidate's dissertations have been defended under his guidance.

Vladimir N. Burkov: doctor of technical sciences, professor, head of the laboratory of active systems at V.A.Trapeznikov Institute for Control Problems at the Russian Academy of Sciences (Moscow), full member of the Russian Academy of Sciences (Moscow), the USSR State Prize Laureat and J.Phon-Neyman Prize Laureat. He has been working in the sphere of controlling large-scale systems for more than 30 years. He is Vice-President of the Russian association for controlling projects, the founder of the theory of active systems, one of the leading specialists in controlling socio-economic systems, including analysis and synthesis of organizational mechanisms.

Ivan A. Gorgidze: doctor of technical sciences, professor, the USSR State Council of Ministers Prize Laureat, one of the first specialists in the theory of active systems control, director of the scientific-research informatization center, academician-secretary of the Georgian Engineering Academy, advisor of the Rector's office of the Georgian Technical University, member of a number of international academies and organizations.

Guram S. Javakhadze: doctor of technical sciences, professor, vice-president of the Engineering Academy of Georgia, head of supervisory board of JSC "Medinserv", one of the leading specialists in the theory of management, member of a number of international academies and organizations.

Ramaz A. Khurodze: doctor of technical sciences, professor, corresponding member of Georgian Academy of Sciences, Georgian State Prize Laureat, Rector of Georgian Technical University, one of the leading specialists in the control theory, member of a number of international academies and organizations.

INDEX

D

E

F

G

Q

R

S

T

U

V

W